Mass Spectrometry-Based Metabolomics

A Practical Guide

Mass Spectrometry-Based Metabolomics

A Practical Guide

Edited by
Sastia Prama Putri
Eiichiro Fukusaki

CRC Press
Taylor & Francis Group
Boca Raton London New York

CRC Press is an imprint of the
Taylor & Francis Group, an **informa** business

CRC Press
Taylor & Francis Group
6000 Broken Sound Parkway NW, Suite 300
Boca Raton, FL 33487-2742

First issued in paperback 2023

Version Date: 20140911

ISBN-13: 978-1-4822-2376-7 (hbk)
ISBN-13: 978-1-03-265224-5 (pbk)
ISBN-13: 978-0-429-16159-9 (ebk)

DOI: 10.1201/b17793

This book contains information obtained from authentic and highly regarded sources. Reasonable efforts have been made to publish reliable data and information, but the author and publisher cannot assume responsibility for the validity of all materials or the consequences of their use. The authors and publishers have attempted to trace the copyright holders of all material reproduced in this publication and apologize to copyright holders if permission to publish in this form has not been obtained. If any copyright material has not been acknowledged please write and let us know so we may rectify in any future reprint.

Publisher's Note
The publisher has gone to great lengths to ensure the quality of this reprint but points out that some imperfections in the original copies may be apparent.

Library of Congress Cataloging-in-Publication Data

Mass spectrometry-based metabolomics : a practical guide / editors, Sastia Prama Putri and Eiichiro Fukusaki.
    p. ; cm.
    Includes bibliographical references and index.
    ISBN 978-1-4822-2376-7 (hardcover : alk. paper)
    I. Putri, Sastia Prama, 1982- editor. II. Fukusaki, Eiichiro, 1960- editor.
    [DNLM: 1. Metabolomics. 2. Mass Spectrometry. QU 120]

    QP171
    612.3'9--dc23   2014023163

Visit the Taylor & Francis Web site at
http://www.taylorandfrancis.com

and the CRC Press Web site at
http://www.crcpress.com

Dedication

*This book is dedicated to
Aisha*

Contents

Preface

In the last decade of the 20th century, biological research has seen a vast accumulation of genomic and transcriptomic information that contributed greatly to the rapid advancement of functional genomics, a field where dynamic aspects resulting from the integration of genomic and transcriptomic knowledge as opposed to the static aspects based on genome information are deemed more substantial and essential. However, this may not be sufficient for the complete elucidation of more complicated quantitative phenotypes. Metabolites, resulting from the interaction of the system's genome with its environment, are not merely the end product of gene expression but also form part of the regulatory system of various biological processes in an integrated manner. Hence, profiling of metabolites is a promising tool for addressing this limitation.

Metabolomics, the global quantitative assessment of metabolites in a biological system, is considered the answer to the problem of analyzing complicated quantitative phenotypes. In addition, integration of metabolomics information into the upstream "omics" such as genomics, transcriptomics, and proteomics would be very important albeit more difficult than expected. Metabolomics involves three different technical elements, namely biology, analytical chemistry, and informatics. Several analytical methods have been utilized for metabolomics and among them, mass spectrometry is one of the most important techniques due to its high sensitivity and quantitative capability. Mass spectrometry-based metabolomics can be a very demanding technique for biologists without special experience in quantitative mass spectrometry, thus the purpose of this book is to explain the complicated know-how of metabolomics. A number of case studies are also introduced for easy understanding of the metabolomics workflow and its practical applications in various research fields. Based on our extensive experience in mass spectrometry-based metabolomics, I decided to summarize our past achievements in the establishment of robust and reproducible protocols from sample preparation to data analysis. This book lays out the fundamental concepts and principles in metabolomics research (Chapter 1) and can act as a good practical guide to start metabolomics research. The whole workflow is also extensively covered in this book that includes experimental design (Chapter 2), preparation of biological samples (Chapter 3), analysis using various instruments (Chapter 4 and Chapter 5), and data processing and data analysis (Chapter 6). In addition, case studies are illustrated in Chapter 7.

Finally, I would like to express my special gratitude to our editors and publisher for providing us with the opportunity of writing this book. I hope that this book will be useful for academicians and researchers of all disciplines in their metabolomics pursuits.

Eiichiro Fukusaki

About the Editors

Dr. Sastia Prama Putri is an assistant professor at the Graduate School of Engineering, Osaka University. She received her PhD from the International Center for Biotechnology, Osaka University, in which she worked on the discovery of novel bioactive compounds from natural products. Her current research includes the application of metabolomics technology for optimizing biofuel production. She is also involved in food metabolomics studies focusing on food authentication and quality evaluation. She is the current chair of the Early Career Members Network (ECMN) of the Metabolomics Society and a board member of the Metabolomics Society.

Professor Eiichiro Fukusaki is a full professor at the Graduate School of Engineering, Osaka University. Previously, he was a deputy chief researcher at Nitto Denko Corporation before returning to an academic career. He has published more than 200 journal articles, book chapters, and reviews. He also holds 25 domestic and 10 international patents. His research collaborators include over 30 academic institutions and major companies from various fields, such as electrical, pharmaceutical, and medical as well as the food industry. He received the Japan "Saito" Award from the Society of Biotechnology Japan in 2004.

Contributors

Dr. Takeshi Bamba
Bioresource Engineering
  Laboratory
Department of Advanced
  Science and Biotechnology
Graduate School of
  Engineering
Osaka University, Japan

Dr. Yusuke Fujieda
Asubio Pharma Co., Ltd.
Kobe, Japan

Prof. Eiichiro Fukusaki
Bioresource Engineering
  Laboratory
Department of Advanced
  Science and Biotechnology
Graduate School of
  Engineering
Osaka University, Japan

Dr. Yoshihiro Izumi
Bioresource Engineering
  Laboratory
Department of Advanced
  Science and Biotechnology
Graduate School of
  Engineering
Osaka University, Japan

Udi Jumhawan
Bioresource Engineering
  Laboratory
Department of Advanced
  Science and Biotechnology
Graduate School of
  Engineering
Osaka University, Japan

Dr. Walter A. Laviña
Bioresource Engineering
  Laboratory
Department of Advanced
  Science and Biotechnology
Graduate School of
  Engineering
Osaka University, Japan

Dr. Arjen Lommen
RIKILT Wageningen UR
Wageningen, The Netherlands

Dr. Fumio Matsuda
Metabolic Engineering
  Laboratory
Department of Bioinformatics
  Engineering
Graduate School of
  Information Science and
  Technology
Osaka University, Japan

Dr. Sastia Prama Putri
Bioresource Engineering
    Laboratory
Department of Advanced
    Science and Biotechnology
Graduate School of
    Engineering
Osaka University, Japan

Dr. Hiroshi Tsugawa
Metabolome Informatics
    Research Team
Metabolomics Research
    Group
RIKEN Center for Sustainable
    Resource Science
Yokohama, Kanagawa, Japan

1
Metabolomics in a Nutshell

Chapter 1

Metabolomics in a Nutshell

Sastia Prama Putri and Eiichiro Fukusaki

Chapter Outline

1.1 Mass Spectrometry-Based Metabolomics

Metabolomics is an interdisciplinary study that involves the comprehensive quantitative profiling of metabolites in a target organism using sophisticated analytical technologies. It is a powerful approach that allows researchers to examine variation in total metabolite profiles, and is capable of detecting complex biological changes using statistical multivariate pattern recognition methods (chemometrics; Putri, et al., 2013b). The main analytical technologies in metabolomic studies are either based on nuclear magnetic resonance (NMR) spectroscopy or mass spectrometry (MS). In addition to these two main technologies, Fourier transform–infrared (FTIR) spectroscopy have also been used. Mass spectrometry requires a preseparation of the metabolic components using either chromatography techniques such as gas chromatography (GC) or liquid chromatography (LC), or other separation techniques such as capillary electrophoresis (CE) coupled to MS (Nicholson, Holmes, and Lindon, 2007).

The main advantage of NMR is its noninvasiveness and non-reliance on analyte separation, thus making the samples recoverable for further analyses. In addition, sample preparation for NMR analysis is relatively simple and easy, and is therefore particularly useful for metabolite profiling of intact biofluids and semisolid or solid samples (e.g., intact cells or tissues). However, the major drawback of NMR is its low sensitivity and the difficulty to annotate the metabolites signals to mine useful biological information (Putri et al., 2013b). Although MS is inherently more sensitive than NMR, it is not possible to recover the samples after analysis by MS. Moreover, it is generally necessary to employ different extraction procedures and separation techniques for different classes of compounds. Despite these disadvantages, MS-based metabolomic studies recently became the mainstream research in this emerging field. The main advantages of MS are high resolution, high sensitivity, a wide dynamic range, coverage of a wide chemical diversity, robustness, and feasibility to elucidate the MW and structure of unknown compounds (Garcia-Cañas, Simó, León, Ibáñez, and Cifuentes, 2011). This book presents a comprehensive overview of the mass spectrometry–based metabolomics, covering the whole pipeline involved in this rapidly emerging field.

1.2 Metabolomics in Functional Genomics and Systems Biology

Metabolomics, the global quantitative assessment of metabolites in a biological system, has played a pivotal role in various fields of science in the postgenomic era. Metabolites, resulting from the interaction of the system's genome with its environment, are not merely the end product of gene expression but also form part of the regulatory system of various biological processes in an integrated manner. Metabolomics complements other "omics" (such as transcriptomics and proteomics) and because it is the "downstream" result of gene expression, changes in the metabolome are considered to be the best reflection of the activities of the cell at a functional level. As one of the omics technologies, metabolomics has exciting applications in various fields, including medical science, synthetic biology, medicine, and predictive modeling of plant, animal, and microbial systems. In addition, integrated applications with genomics, transcriptomics, and proteomics provide greater understanding of global systems biology (Putri et al., 2013a). Compared to the other omics fields, the tools used in metabolomics are still constantly developing. Development of new analytical methods with the aim of analyzing as many metabolites in a single run and achieving faster analysis, optimization of sample preparation protocols, and the establishment of mass spectral libraries are several examples of the key topics in metabolomics. The unification and standardization for conducting metabolomics experiments are also an ongoing effort in the metabolomics community.

1.3 Power of Metabolomics for Systems Biology: Informative, Predictive, Discriminative

The power of metabolomics lies in the acquisition of analytical data in which metabolites in a cellular system are quantified, and the extraction of the most meaningful elements of the data by using various data analysis tools. The field of metabolomics has continued to grow rapidly over the last decade and has been proven to be a powerful technology in predicting and explaining complex phenotypes in diverse biological systems. Based on the specific objective of the analysis, applications of metabolomics technology in systems biology include (i) informative analyses to characterize and identify compounds of interest, (ii) prediction of various phenotypes by means of multivariate analysis using metabolome data as the explanatory variable (herein termed "predictive metabolomics"), and (iii) comparative metabolomics to determine the metabolites responsible for classification of samples by type or for discriminatory purposes (Putri, Bamba and Fukusaki, 2014, Cevallos-Cevallos, Reyes-De-Corcuera, Etxeberria, Danyluk, and Rodrick, 2009).

In general, the main framework of discriminative metabolomic study is to: (i) conduct metabolic profiling or fingerprinting by an analytical instrument(s), (ii) visualize the data structure and identify factors that enable sample classification using multivariate analysis, and (iii) identify biomarkers for classification. On the other hand, the main framework of predictive metabolomic study is to: (i) conduct metabolic profiling or fingerprinting by a choice of analytical instrument and multivariate analysis, (ii) create a predictive model that corresponds to the phenotype of a given organism or biological sample from the metabolite data set, and (iii) classify the component groups that contribute to the predictive model. Informative metabolomics study usually focused on the identification and quantification of targeted or untargeted metabolites to obtain sample intrinsic information. An excellent example of informative metabolomics is the development and continuous update of metabolite databases such as the human metabolome database (Wishart et al., 2007).

1.4 Practical Workflow of Metabolome Analysis

In general, the metabolomics workflow includes experimental design (Chapter 2), sample preparation of biological samples (Chapter 3), metabolite analysis using various instruments (Chapter 4 and Chapter 5), and data processing and data analysis (Chapter 6) as seen in Figure 1.1. In order to perform metabolomics research, researchers require skills from three main research disciplines: bioscience, analytical chemistry, and informatics. These interdisciplinary disciplines are essential to achieve successful metabolomics work.

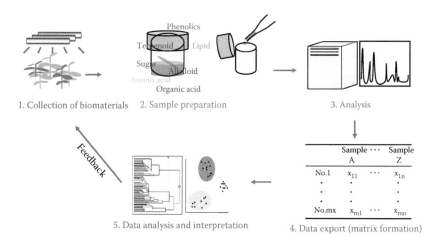

Figure 1.1 Metabolomics workflow. (Modified from Putri, Bamba, and Fukusaki, 2014, with permission from Future Science Ltd.)

1.5 Current Challenges and Future Prospects in Metabolomics Studies

Annotation of metabolite signals is one of the persisting problems that need to be resolved for further progression of metabolomics research. At present, identification of unknown metabolites is the major bottleneck in the field. Although hundreds to thousands of metabolites can be detected in a single experiment, most of them remain unidentified due to limited metabolite information in the library and unavailability of reference standards. Several technological challenges also need to be addressed for a wider implementation of metabolomics. Further developments in analytical instrumentation are desired, particularly to increase the dynamic range in order to reliably quantify metabolites of extremely low to very high abundance, and to widen the coverage of the metabolome by enabling the analysis of metabolites with vastly different chemical properties. Rapid development in analytical instrumentation poses another challenge for the development of more powerful data processing. Further development of bioinformatic resources to facilitate easy data interpretation from complex metabolomics data sets is also required.

References

Cevallos-Cevallos, J. M., Reyes-De-Corcuera, J. I., Etxeberria, E., Danyluk, M. D., and Rodrick, G. E., Metabolomic analysis in food science: A review, *Trends in Food Science Technology,* 2009, 20: 557–566.

García-Cañas, V., Simó, V., León, Carlos, Ibáñez, E., and Cifuentes, A., MS-based analytical methodologies to characterize genetically modified crops, *Mass Spectrometry Reviews*, 2011, 30: 396–416.

Nicholson, J. K., Holmes, E., and Lindon, J. C., Metabonomics and metabolomics techniques and their applications in mammalian systems, in J. C, Lindon, J. K, Nicholson, and E. Holmes (Eds.), *The Handbook of Metabonomics and Metabolomics*, 2007, Oxford: Elsevier.

Putri, S. P,, Bamba, T., and Fukusaki, E., Application of metabolomics for discrimination and sensory predictive modeling of food products, in *Hot Topics in Metabolomics: Food and Nutrition* (e-book). Future Science, London, UK. 2014.

Putri, S. P., Nakayama, Y., Matsuda, F., Uchikata, T., Kobayashi, S., Matsubara, A., and Fukusaki, E., Current metabolomics: Practical applications, *Journal of Bioscience and Bioengineering*, 2013a, Epub ahead of print. doi: 10.1016/j.jbiosc.2012.12.007.

Putri, S. P., Yamamoto, S., Tsugawa, H., and Fukusaki, E., Current metabolomics: Technological advances, *Journal of Bioscience and Bioengineering*, 2013b, Epub ahead of print. doi:10.1016/j. jbiosc.2013.01.004.

Wishart, D. S., Tzur, D., Knox, C., Eisner, R., Guo, A. C., Young, N., et al., HMDB: The human metabolome database, *Nucleic Acids Research*, 2007, 35: 521–526.

2
Design of Metabolomics Experiment

Chapter 2

Design of Metabolomics Experiment

Sastia Prama Putri, Fumio Matsuda, and Takeshi Bamba

Chapter Outline

2.1 Introduction

Experimental design in metabolomics studies includes five main aspects: (1) generation of a working hypothesis, (2) determination of metabolomics approach, (3) data acquisition (e.g., analytical conditions), (4) sample preparation (preparation of biological samples, quenching, and extraction methods), and (5) data analysis (use of statistical analyses to visualize the data). Although the importance of these topics is relevant to all biological analyses, there are several points that require special consideration for a successful metabolomics work. In this chapter, several technical tips and know-how are introduced and discussed.

2.2 Start with a Working Hypothesis

A metabolomics study should begin with asking the right biological question. Once the problem that requires an answer is fully understood, a suitable working hypothesis can be generated. A working hypothesis is a premise that serves as a basis for further investigation. An appropriate research strategy is then formulated to test the working hypothesis. In order to develop a solid working hypothesis, researchers initially need to collect a great amount of background information about the organism of choice. This will not only help in the data interpretation but also in making a good experimental design so as to understand how many and what kind of samples are really important for analysis. Upon generation of a working hypothesis, the researcher can then decide on the type of analysis (targeted or nontargeted), appropriate analytical tools, and the method for biological sample preparation. Once the metabolome data are acquired, data mining using either univariate or multivariate statistical tools can be performed. The subsequent steps include interpretation of the findings, generation of new hypotheses, and further verification of the new hypotheses by improved experimental design.[1]

2.3 Design of Analysis: Targeted or Nontargeted Approach?

There are three major approaches used in metabolomics: (1) targeted analysis (precise and quantitative measurement of the concentration of a limited number of known metabolites), (2) metabolic profiling (untargeted high-throughput measurement of the levels of a large number of metabolites, including

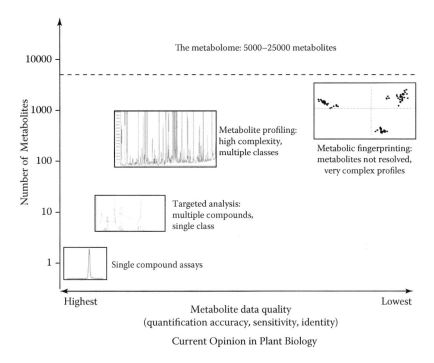

Figure 2.1 Graphical representation of the trade-off between metabolite coverage and the quality of metabolic analysis. (From Trethewey, R. N., *Current Opinion in Plant Biology*, 2004, 7: 196–201, with permission from Elsevier.)

unknown metabolites), and (3) metabolic fingerprinting (rapid and total evaluation of biochemical fingerprints for discrimination of different groups in which metabolite identification is not required).[2] As depicted in Figure 2.1, there is a trade-off between metabolic coverage and the quality of metabolic analysis. Depending on the purpose of the experiment, one should decide whether the important point to consider is data quality (i.e., quantification accuracy, sensitivity, and identity), or coverage of metabolites detected. The higher the number of metabolites included in the analysis, the lower the metabolome data quality becomes.[3] It is therefore important to decide the type of metabolomics approach to pursue beforehand. After deciding the type of analysis and which metabolites will provide substantial information, the researcher can then move on to deciding the appropriate analytical instrument to be used.

2.4 Deciding the Appropriate Analytical Instrument[4]

The goal of metabolomics is the exhaustive profiling of all metabolites in biological samples. However, because metabolites vary greatly (e.g., in molecular weight, polarity, solubility, etc.), a method capable of analyzing all metabolites simultaneously is currently unavailable. Therefore, it is often necessary to narrow down the number of target metabolites and acquire the metabolome data using appropriate instruments or combine data from several instruments. Here, we describe the development, characteristics, and the latest applications of the analytical instruments that are commonly used in metabolomics with special emphasis on the development of mass spectrometry because it has the most impact in the recent growth of metabolomics.

2.4.1 Nuclear Magnetic Resonance (NMR)

NMR has been a commonly used analytical tool for metabolomics research for the last decade. This method detects the specific resonance absorption profiles of metabolites in a magnetic field (which is dependent on chemical structure). One major advantage of NMR analysis is its noninvasiveness and nonreliance on analyte separation, thus the samples can be recovered for further analyses. Therefore, after NMR analysis, the samples can be further analyzed by gas chromatography/mass spectrometry (GC/MS) and liquid chromatography/mass spectrometry (LC/MS). In addition, sample preparation for NMR analysis is relatively simple and easy, and therefore it is particularly useful for metabolite profiling of intact biofluids[5] and semisolid or solid samples (e.g., intact cells or tissues).[6] Furthermore, NMR can be utilized for the analysis of food samples that are composed of diverse compound mixtures, because signal intensity is less affected by other components in the sample. Despite being a superior instrument with regard to sample recovery and ease of use, the major drawback of NMR is its low sensitivity. However, the use of labeled compounds can lead to significant improvements in sensitivity (e.g., using ^{13}C or ^{15}N NMR).

One recent application of NMR-based metabolomics is the study of Kikuchi, Shinozaki, and Hirayama using ^{13}C and ^{15}N-labeled *Arabidopsis thaliana*.[7] They analyzed [1]H-^{13}C heteronuclear single-quantum correlation spectral differences to study changes in metabolite content between wild type and mutant strains.

2.4.2 GC/MS

One of the most successful methods in metabolomics is the GC/MS, a method combining gas chromatography and mass spectrometry, developed by Oliver Fiehn in 2000.[8] There is a large amount of standard compound information for metabolite identification for GC/MS. This is due to the high repeatability of peak retention time by the introduction of the retention index (RI). Because of this advantage, the GC/MS method has been used for more than 10 years with only minor modifications to the detection method, such as the introduction of time-of-flight (TOF) and triple quadruple (QqQ)/MS. GC/MS requires derivatization of samples for nonvolatile target compounds, because these have to be heated to the gas state. Low-molecular–weight hydrophilic metabolites (e.g., sugars, organic acids, amino acids, etc.) can also be analyzed by combining two derivatization methods. Methoxyamine hydrochloride and *N*-methyl-*N*-(trimethylsilyl)trifluoroacetamide, are used for the first and second derivatizations, respectively.[9] If the target metabolites are high-molecular–weight compounds that are not volatilized even after derivatization, it is necessary to complement the analysis with other instruments. This is also the case for compounds that are not resistant to degradation by heat.

Compound identification using GC/MS is relatively easy compared with other analytical platforms because the unique mass spectrum of each compound can be consistently obtained. This is achieved through the use of electron ionization (EI), which is currently the most commonly used ionization method in GC/MS owing to its robustness and high repeatability. Furthermore, a large number of compounds can be easily identified by using the National Institute of Standards and

Technology mass spectral library. The accuracy of peak identification has been significantly improved with the recent creation of a semiautomated peak identification system.[10]

One of the latest advancements in the GC system is the development of a two-dimensional gas chromatography/mass spectrometry system (GC × GC/MS). This system provides a significant increase in peak capacity, as well as resolution and sensitivity, compared with GC/MS.[11] Waldhier et al. have used GC × GC/MS to improve the analytical resolution of amino acid enantiomers in serum and urine.[12]

2.4.3 LC/MS

LC/MS is a combination of liquid chromatography and mass spectrometry. Analysis of a wide range of metabolites, ranging from high- to low-molecular–weight, and from hydrophilic to hydrophobic, can be performed by selecting the appropriate column and mobile phases. Electrospray ionization (ESI), a general ionization method in LC/MS, can ionize most compounds.[13] However, ESI lacks quantitative capability due to the occurrence of a phenomenon first described by Ikonomou, Blades, and Kebarle[14] called the *matrix effect* (ionization suppression and enhancement due to the coeluted matrix). Labeling compounds with stable isotopes and spiking them into the sample is one way to avoid this matrix effect.[15]

In the past, LC/MS was considered unsatisfactory for the analysis of hydrophilic metabolites. However, in recent years, simultaneous analysis of hydrophilic metabolites (amino acids, organic acids, nucleic acids, etc.) has been achieved using pentafluorophenylpropyl high-performance liquid chromatography (HPLC) stationary phases without any derivatization.[16] In addition, sugars, sugar alcohols, and sugar phosphates have been successfully analyzed using ZIC®-HILIC column separation, a type of hydrophilic interaction chromatography (HILIC).[17] Moreover, reversed-phase ion-pair LC is also a useful method for metabolomics.[18] This system can detect negatively charged compounds including sugar phosphates, nucleotides, and carboxylic acids, which are involved in the central carbon metabolism (including glycolysis, pentose phosphate pathway, and tricarboxylic acid cycle). An ion-pair reagent is a compound that has a countercharge compared with the target metabolites in water solution. Although the use of an ion pair reagent is very useful, the most serious problem is the matrix effect in ESI-MS based on coelution of impurities from the sample or ion-pair reagent from the mobile phase. Caution should be exercised when using an ion-pair reagent because remaining unpaired reagent contaminating the ESI-sprayer or inside the MS may affect the sensitivity or repeatability of LC/MS.

The release of the ultra-performance liquid chromatography (UPLC®) instrument (Waters Corporation, United States) in 2004 has played a big part in the advancement of LC/MS-based metabolomics.[19] Significant improvements provided by the UPLC system include the use of a high-pressure resistant column with a 1.7-μm carrier and a pump with a maximum pressure of 100 MPa. This results in shorter measurement time, improved sensitivity, and higher peak capacity. In tandem with UPLC, several companies, including Shimadzu Corporation (Kyoto, Japan) and Agilent Technologies (Palo Alto, CA), and Thermo Fisher Dionex (Sunnyvale, CA) developed ultra-high–performance liquid chromatography (UHPLC), which can be used with pressures exceeding 100 MPa.

Despite these significant improvements in liquid chromatography techniques, peak capacity and stability are still not sufficient for simultaneous analysis of hydrophilic metabolites. There is still room for improvement in this respective research area.

In the case of LC/MS metabolomics, there is no method to standardize retention time information, such as the RI commonly used in GC/MS. Thus, the development of an analytical method to ensure a constant retention time across the whole chromatogram is required. High repeatability of the retention time, data processing to generate a data matrix as well as a collection of retention time information for many standard compounds should be attained. The stability of each component in the data acquisition procedure is also important for construction of a robust method for large-scale metabolome analysis.

2.4.4 CE/MS

Capillary electrophoresis mass spectrometry (CE/MS) was used for metabolomics by the group of Soga and Heiger in 2000.[20]

They developed a method for amino acid detection without derivatization, as well as a system for analysis of intermediates of the glycolytic system, pentose phosphate pathway, and citric acid cycle. Using this system, carboxylic acids, phosphorylated carboxylic acids, phosphorylated saccharides, nucleotides, and nicotinamide and flavin adenine coenzymes of glycolysis and the tricarboxylic acid cycle pathways were also successfully analyzed. To analyze these anionic metabolites, electroosmotic flow that moves from cathode to anode is necessary, thus they used a specially coated (SMILE(+)) capillary to address this problem.[21] However, multivalent ions such as nucleotides and CoA are adsorbed onto the surface of the internal wall of the SMILE(+) capillary, thus an alternative pressure-assisted CE/MS technique using a DB1 capillary that is a noncharged polymer-coated capillary was recently developed.[22] CE/MS is superior compared to HPLC/MS in terms of separation capacity. Although ion and ion-pair chromatography have been reported as suitable methods for comprehensive analysis of ionic compounds,[15] CE/MS affects MS less compared with these methods. However, low repeatability is a major disadvantage of CE/MS, which is mainly caused by variation in migration time as it is easily influenced by temperature changes.

As the method developed by Soga et al. required a combination of two methods for the study of anion metabolites.[21,22] Harada et al.[23] developed a more comprehensive analytical method for anion metabolites. In conventional CE analysis, the inlet of the capillary is set at the anode and the outlet is at the cathode. In this new method, however, the CE polarity for anions is inverted relative to the conventional CE setup. Moreover, an ordinary fused silica capillary was chosen instead of a specific cationic polymer-coated capillary (SMILE (+) capillary). Due to the high electrolyte pH, electroosmotic flow exceeds the electrophoretic mobility of all target metabolites. Furthermore, this method overcomes adsorption of anionic metabolites by the capillary because the fused silica capillary surface has the negative charge; hence, most of the anions migrate toward the MS simultaneously.[23] Using a sulfonated capillary and multiple reaction monitoring (MRM), reproducibility and sensitivity were improved and analysis time was reduced.[24] Subsequently, Soga et al. examined the materials constituting the electrosprayer, where they found that sensitivity is improved by using platinum.[25] One example of CE/MS-based metabolomics is the metabolome analysis of saliva (obtained from 215 individuals) performed by Sugimoto and coworkers in which they demonstrated that salivary metabolomics can be an alternative method for the detection of oral, breast, and pancreatic cancers.[26]

2.4.5 SFC/MS

Although hydrophilic metabolites are common analytical targets in metabolomics, lipids have come into the spotlight as an attractive target because recent research revealed that lipids act as signaling molecules in various biological phenomena.[27] Bamba et al. developed a new method for the analysis of lipids using supercritical fluid chromatography mass spectrometry (SFC/MS).[28] SFC, which has characteristics of both GC and HPLC, is a high-resolution and high-throughput separation method using supercritical fluid (noncondensable fluid over liquid-gas at the critical point) as a mobile phase. Due to the high diffusion coefficient of the supercritical fluid, the separation capacity of SFC is much higher compared to HPLC. In addition, inasmuch as the polarity of the supercritical fluid carbon dioxide is similar to hexane, SFC can also be employed for the analysis of hydrophobic metabolites. Moreover, the polarity of the mobile phase can be significantly altered by the addition of a polar organic solvent (such as methanol) as a modifier. Moreover, a wide range of separation modes that are not available in GC and HPLC can be selected by controlling temperature or back pressure. Thus, SFC/MS can be used for the high-throughput and high-resolution analysis of diverse metabolites.[29]

Recently, a lipidomics system using SFC/MS has been constructed for the analysis of various types of lipids, including phospholipids, acylglycerols, cholesterols, and oxidized phospholipids.[30,31] In addition, Matsubara, Bamba, Ishida, Fukusaki, and Hirata established a high-speed and high-resolution profiling method for a carotenoid mixture using SFC/MS.[32] Aside from SFC/MS, SFC can also be combined with supercritical fluid extraction (SFE) online. This so-called online SFE-SFC/MS system can be effective for unstable metabolites in a typical solvent extraction method. For example, SFC/SFE has been used for the analysis of both reduced and oxidized forms of Coenzyme Q_{10}.[33]

2.4.6 MALDI-MS

Matrix-assisted laser desorption/ionization (MALDI) is an ionization technique using an excess amount of matrix that is uniformly dispersed with the sample. In this technique, the sample is ionized by nitrogen laser pulse irradiation (i.e., ultraviolet) to the surface of the mixture and the matrix has three primary roles: (1) assisting ionization of target compounds by transmitting laser energy, (2) suppression of photodecomposition of targets, and (3) dissociation of the molecular ion or specific binding by suppression or enhancement of fragmentation. MALDI-MS provides low selectivity because, unlike LC/MS, the analyte is not separated by chromatography. Although time-of-flight MSs have been used in MALDI-MS,[34] it is not sufficient to improve the selectivity of MALDI-MS significantly. Therefore, other MS methods, such as Fourier transform ion cyclotron resonance (FT-ICR), orbitrap, or TOF/TOF-type MS, have recently been used to enhance the selectivity of MALDI-MS.[35,36]

Due to poor quantitative performance, low repeatability, and interference of ionization by the matrix, MALDI-MS has been used primarily for the identification of high-molecular–weight molecules such as proteins. A recent report using this system to analyze a mixture of 30 low-molecular–weight metabolites demonstrated that ionization suppression at high concentration was the main factor requiring further improvement.[37] Exciting applications of MALDI-MS include the development of a highly sensitive technique for nontargeted metabolomics that would be practically useful for the analysis of cellular metabolites.[38] Yukihira, Miura, Saito, Takahashi, and Wariishi developed a method for the analysis of intracellular metabolic dynamics.[39] In addition, by applying this MALDI-MS method, Miura et al. developed an ultrasensitive *in situ* metabolomics imaging method that can visualize metabolites in two dimensions.[40] This technique is considered an innovative multimolecular imaging technique that should be useful for compound analysis in pathological conditions.

2.4.7 Direct Infusion MS

In direct infusion MS (DI-MS), the sample is directly introduced into the ESI source without chromatographic separation by using a syringe pump or nanospray chip. DI-MS is a very high throughput method due to its short analysis time. However, because the matrix effect is quite strong, the quantitative performance is inferior to LC/MS. Although stable isotope labeling has been used to overcome the matrix effect,[41] the cost of this method is too high for multiple component analysis. In addition, the molecular selectivity of DI-MS is inferior to LC/MS and MALDI-MS due to a lack of retention time information for the metabolites. To address this limitation, MS/MS and accurate mass analysis have been commonly applied to DI-MS metabolomics by FT-ICR and orbitrap MS.[42] Furthermore, improvement on the dynamic range and mass accuracy of direct infusion nanoelectrospray Fourier transform ion cyclotron resonance mass spectrometry (DI-nESI FT-ICRMS) was achieved by the spectral stitching method. This method comprises the collection of multiple adjacent selected ion-monitoring (SIM) windows that are stitched together using novel algorithms.[43]

The use of flow injection analysis/mass spectrometry (FIA/MS), an automated method using HPLC without a column, is gaining popularity. FIA/MS analysis has been used for metabolite fingerprinting,[44] screening in drug discovery,[45] and pesticide multiresidue detection in food.[46] For example, Fuhrer, Heer, Begemann, and Zamboni developed a platform that can analyze 1,400 *Escherichia coli* cellular extract samples per day.[47] However, this method has to be combined with chromatography to expand the coverage of target metabolites.

One of the most important developments in DI-MS is the use of the NanoMate chip-based nanoelectrospray system.[48] It is a fully automated system that uses a new sample delivery tip and electrospray nozzle for each analysis to enable high-throughput analysis with no cross-contamination.[43] One recent application of this system is for shotgun lipidomics using the LTQ Orbitrap XL instrument (Thermo Fisher Scientific, Bremen, Germany) equipped with a robotic nanoflow ion source, TriVersa NanoMate (Advion BioSciences, Ithaca, NY).[49]

At present, because there is no universal instrument capable of measuring every type of metabolite, users should select the most appropriate method for their research, giving consideration to several factors, including resolution, sensitivity, throughput, and cost.

2.5 Design of Sample Preparation Procedure and Importance of Method Validation

Although preparation of biological samples is not a techno-logical issue, it is the most important factor for a successful metabolomics study because a metabolome analysis using advanced, smart, and developed technologies will not produce a relevant result from poorly prepared biological samples. This implies that successful metabolomics studies are based on the good communication between the analytical chemist and the biologist. It should be noted that, although there are limited time and resources available for metabolome analysis in the analytical chemist's lab, the number of analytical samples requested by a biologist tends to increase in a geometric series. For example, the requested sample for a phenotyping experiment of a single mutant often numbers over 100 (144 samples are needed for a triplicate analysis of four genotypes: wild type, two alleles, one revertant) in three different organs at four different time points). From our experience, productive metabolome analyses with good biological results have been performed with less than 30 samples. In this case, inasmuch as the maximum number of samples that can be analyzed in one batch is 30, it is important for the biologist to have enough background data in order to reduce the number of samples that best represent the experiment. This highlights the impor-tance of understanding the biological question that needs to be answered, which subsequently can result in a highly effi-cient experimental design.

All analytical analyses, including metabolomics studies, have to be validated in terms of accuracy, precision, repeatability, intermediate precision, specificity, and detection limit before being applied to biological analyses. Although the introduction of quality control (QC) for samples has recently been discussed to control the intermediate precision or intralaboratory repeatability of the analysis,[50,51] there is no established guideline for validation or QC of the metabolome analysis procedure. Practically, the internal standard method has been widely used in metabolome analyses to ensure repeatability on daily analyses. Because the sensitivity of mass spectrometry depends on the setup conditions, a compound of known concentrations is added to the sample extraction buffer in order to cancel the sensitivity drift. The signal intensity data in each analysis can be normalized against the signal intensity data of the standard.[52] Thus, internal standards should have the following characteristics to qualify as usable for metabolome analysis:

(1) It should hardly exist in the biological samples.
(2) It should not seriously overlap with metabolite signals derived from the samples.
(3) It should not be affected by ion suppression in procedures using electrospray ionization.
(4) It should be inexpensive and stable in stock solution.

Based on the above criteria, several useful intermediates have been selected for metabolome analyses using GC/MS (such as ribitol, adipate, and isotope-labeled alanine), LC/MS (d-camphor sulfonate and lidocaine), and CE/MS (methionine sulfone). These internal standards are also used for correction of retention time shifts among analyses.

A rapid quenching method should be validated, especially for an analysis of metabolites and cofactors related to central metabolism such as glucose phosphates, NAD(P)H, and ATP. Because the pool size of these metabolites is dynamically changed by external stimuli, the enzymes responsible for the metabolism have to be stopped as soon as possible after sampling. Quenching is usually achieved by rapid freezing of the tissues in liquid nitrogen, just after excision of the samples. However, it has been reported that the procedure is too slow to capture an intact metabolic state in samples of microorganisms and tissues of animals.[53-55] There is no perfect method for sample quenching, therefore it is recommended to validate every quenching and metabolite extraction procedure by a suitable method. The procedure for metabolite extraction, partial cleanup, and derivatization, should also be optimized for each biological sample by means of a recovery test using a representative metabolite. The whole procedure can be optimized by employing an efficient methodology for the design of the experiment.[56] More comprehensive information on sample preparation is described in Chapter 3.

2.6 Choosing the Appropriate Data Analysis Method[*]

Data analysis, which aims to find significant changes and validate the data obtained from the biological samples, is a crucial process in metabolomics research. Bioinformatics tools allow researchers to perform data analysis easily, including organized data matrix construction and statistical analysis. Data analysis strategies are usually divided into nontargeted and targeted approaches. A general strategy of data analysis can be found in Figure 2.2.

In the nontargeted approach, the most important task is to construct an organized two-dimensional data matrix consisting of metabolites and quantitative variables from a large amount of raw data sets. A summary of freely available and useful software suited to achieve this goal can found in Table 2.1. Upon construction of the data matrix, the next step is to mine for significant metabolites. The data matrix is usually extremely large, therefore users need to focus on interesting and meaningful metabolites from an organized data matrix. Multivariable analysis is a convenient way to extract the metabolites of interest from such data matrices. A

[*] Content modified from *Journal of Bioscience and Bioengineering*, 2013, 116: 9–16, with permission from Elsevier, license number 3281751226702.

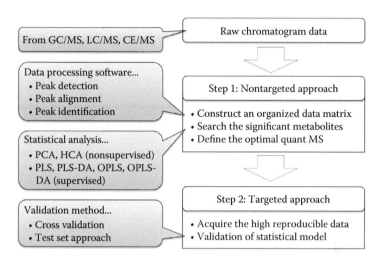

Figure 2.2 Data analysis strategy. (From Putri, S. P., Yamamoto, S., Tsugawa, H., and Fukusaki, E., *Journal of Bioscience and Bioengineering,* 2013, 116: 9–16 with permission from Elsevier.)

summary of common terms in multivariable analysis is shown in Table 2.2. More details on various multivariable analyses can be found in Chapter 6.

After the researcher identifies the metabolites of interest by means of nontargeted analysis and statistical analysis, the next step is the validation of the hypothesis or the statistical model by means of a targeted approach. As opposed to the nontargeted approach, the targeted approach focuses on only the metabolites of interest. Because metabolites obtained using the nontargeted approach can potentially be false-positives or false-negatives due to the analysis procedure (i.e., the approach and accuracy of peak annotation is significantly different from targeted analysis), the researcher should confirm their significance and reliability. In addition, in targeted analysis, it is critical to obtain a highly reproducible data matrix for comparison with previous experiments; thus, development of a quantification method is important.

Finally, it is an important task to get the biological interpretation from the multivariate analysis result. However, the selection of important metabolites and the final biological

TABLE 2.1
Freely Available Software for Nontargeted Approach

Software	Analysis Type	Feature
AIoutput (8)	GC/MS Nontargeted	The mass spectra integration and a data matrix construction are performed from the MetAlign result. Peak identification and prediction are executable based on the user-defined reference library and multivariable analysis.
JDAMP (57)	CE/MS Nontargeted	Differences between metabolite profiles from CE/MS data set are highlighted by applying arithmetic operation to all corresponding signal intensities from whole raw data sets on a datapoint-by-datapoint basis.
Metab (54)	GC/MS Nontargeted	Peak alignment is performed based on the deconvoluted files from AMDIS software in command line style.
MetaboliteDetector (51)	GC/MS Nontargeted and Targeted	Improved AMDIS method is used for deconvolution processing. An organized data matrix is constructed based on MSTs. Peak identification is performed by the user-defined reference library.
MetAlign (52)	GC/MS, LC/MS Nontargeted	Data preprocessing including baseline correction, peak detection, and peak alignment is executable. The basis of nearly all algorithms is derived from the way a trained expert would analyze the data by eye and hand.
MetaQuant (69)	GC/MS Targeted	The metabolite quantification for GC/MS data is executable by a user-defined library.

TABLE 2.1 (CONTINUED)
Freely Available Software for Nontargeted Approach

Software	Analysis Type	Feature
MET-IDEA (70)	GC/MS Targeted	If users prepare an input list of ion/retention time pairs, it is possible to extract the targeted ion chromatographic peak areas and determine the relative metabolite abundances.
MSFACTs (42)	GC/MS Nontargeted	Peak alignment is performed based on the RTAlign tool from the ASCII format data set.
Mzmine (43)	LC/MS Nontargeted	All analysis from raw data to statistical analysis is executable by many data processing methods and graphical user interfaces. Some kinds of identification methods are available.
TagFinder (49)	GC/MS Nontargeted	RI calculation and MSTs extraction are executable for GC/MS data set. Peak identification method utilizes the Golm metabolome database.
XCMS (45)	LC/MS Nontargeted	A nonlinear retention time alignment is utilized especially for the LC/MS data set. Recently, XCMS online has been released to perform a data matrix construction easily without the console input. Peak annotation is executable based on METLIN database.

Source: From Trethewey, R.N., *Current Opinion in Plant Biology*, 2004, 7: 196–201, with permission from Elsevier.

interpretation is often subject to the background knowledge or biases of individual researchers. Metabolite set enrichment analysis (MSEA) is a useful tool to resolve these issues.[57] The idea of MSEA is based on gene set enrichment analysis (GSEA), a widely used technique in transcriptomic data analysis that uses a database of predefined gene sets to rank lists

TABLE 2.2
Common Terms in Multivariable Analysis

Term	Explanation
PCA	Constructing the informative axis (principal component) on the basis of the data variance.
Score	Score (Point) of the data projected on the informative axis. The plot that is usually shown by two or three dimensions is used to understand the features of samples.
Loading	Index of the important degree of variables to construct the informative axis.
Contribution ratio	Amount of information explained by the principal component.
HCA	Clustering the samples or variables on the basis of distance in multidimensional space.
Distance computation	The distance is calculated by Minkowski metric.
Clustering algorithm	Single, Complete, Average, Centroid, and Ward linkage are frequently used. Researchers should select the optimal method based on the situation.
PLS, OPLS	Constructing the informative axis correlated to the supervised variables.
Q2	Guidepost to determine the optimal number of latent variables.
RMSEE	Index from internal validation to evaluate the accuracy and precision of the model.
RMSEP	Index from external validation to evaluate model robustness.
VIP	Criteria for selecting important variables of the PLS model on the basis of the squared PLS weight.
Coefficient	X coefficient for regression of Y.
PLS-DA, OPLS-DA	Discriminant analysis by means of PLS algorithm using 0 and 1 as supervised.
ROC curve	Evaluating the classifier performance on the basis of area under the curve, and setting the optimal cutoff value to discriminate the samples.

Source: From *Journal of Bioscience and Bioengineering*, 2013, 116: 9–16 with permission from Elsevier.

from microarray studies.[58] MSEA also utilizes the predefined metabolite set library to help researchers identify and interpret patterns of living organisms' metabolite changes in a biologically meaningful context. There is some software for MSEA such as Metaboanalyst and MBRole that mainly utilize the overrepresentation analysis or quantitative enrichment analysis to understand objectively what is going on in the living organism.[59,60]

Bioinformatics has been rapidly advancing due to the progress in automated data processing and, consequently, researchers can spend more time interpreting the results and designing future projects. Moreover, if quality assessment and medical diagnostics can be automatically and routinely performed using raw chromatogram data, metabolomics could be extremely useful to the biomedical field. However, users should not casually consider all results relevant inasmuch as complete dependence on automated procedures can lead to critical errors, therefore, manual validation of the output should necessarily be performed.

References

1. Fiehn, O. and Kind, T., Metabolite profiling in blood plasma. In Wolfram Weckwerth (Ed.), *Methods in Molecular Biology: Metabolomics Methods and Protocols*, 2007, Totowa, NJ: Humana Press.
2. Blekherman, G., Laubenbacher, R., Cortes, D. F., Mendes, P., Torti, F. M., Akman, S., Torti, S. V., and Shulaev, V., Bioinformatics tools for cancer metabolomics, *Metabolomics*, 2011, 7: 329–343.
3. Trethewey, R. N., Metabolite profiling as an aid to metabolic engineering in plants, *Current Opinions in Plant Biology*, 2004, 7: 196–201.
4. Putri, S. P., Yamamoto, S., Tsugawa, H., and Fukusaki, E., Current metabolomics: Technological advances, *Journal of Bioscience and Bioengineering*, 2013, 116: 9–16.
5. Nakanishi, Y., Fukuda, S., Chikayama, E., Kimura, Y., Ohno, H., and Kikuchi, J. Dynamic omics approach identifies nutrition-mediated microbial interactions, *Journal of Proteome Research*, 2011, 10: 824–836.

6. Andronesi, O. C., Blekas, K. D., Mintzopoulos, D., Astrakas, L., Black, P. M., and Tzika, A. A. Molecular classification of brain tumor biopsies using solid-state magic angle spinning proton magnetic resonance spectroscopy and robust classifiers, *International Journal of Oncology*, 2008, 33: 1017–1025.

7. Kikuchi, J., Shinozaki, K., and Hirayama, T., Stable isotope labeling of *Arabidopsis thaliana* for an NMR-based metabolomics approach, *Plant Cell Physiology*, 2004, 45: 1099–1104.

8. Fiehn, O., Kopka, J., Trethewey, R. N., Willmitzer, L., Identification of uncommon plant metabolites based on calculation of elemental compositions using gas chromatography and quadrupole mass spectrometry, *Analytical Chemistry*, 2000, 72: 3573–3580.

9. Pongsuwan, W., Fukusaki, E., Bamba, T., Yonetani, T., Yamahara, T., and Kobayashi, A., Prediction of Japanese green tea ranking by gas chromatography/mass spectrometry-based hydrophilic metabolite fingerprinting, *Journal of Agricultural and Food Chemistry*, 2007, 55: 231–236.

10. Tsugawa, H., Tsujimoto, Y., Arita, M., Bamba, T., and Fukusaki, E., GC/MS based metabolomics: Development of a data mining system for metabolite identification by using soft independent modeling of class analogy (SIMCA), *BMC Bioinformatics*, 2011, 12: 131.

11. Lee, A. L., Bartle, K. D., and Lewis, A. C., A model of peak amplitude enhancement in orthogonal two-dimensional gas chromatography, *Analytical Chemistry*, 2001, 73: 1330–1335.

12. Waldhier, M. C., Almstetter, M. F., Nürnberger, N., Gruber, M. A., Dettmer, K., and Oefner, P. J., Improved enantiomer resolution and quantification of free D-amino acids in serum and urine by comprehensive two-dimensional gas chromatography-time-of-flight mass spectrometry, *Journal of Chromatography A*, 2011, 1218: 4537–4544.

13. Whitehouse, C. M., Dreyer, R. N., Yamashita, M., and Fenn, J. B., Electrospray interface for liquid chromatographs and mass spectrometers, *Analytical Chemistry*, 1985, 57: 675–679.

14. Ikonomou, M. G., Blades, A. T., and Kebarle, P., Investigations of the electrospray interface for liquid chromatography/mass spectrometry, *Analytical Chemistry*, 1990, 62: 957–967.

15. Izumi, Y., Okazawa, A., Bamba, T., Kobayashi, A., and Fukusaki, E., Development of a method for comprehensive and quantitative analysis of plant hormones by highly sensitive nanoflow liquid chromatography-electrospray ionization-ion trap mass spectrometry, *Analytica Chimica Acta*, 2009, 26: 215–225.

16. Yoshida, H., Mizukoshi, T., Hirayama, K., and Miyano, H., Comprehensive analytical method for the determination of hydrophilic metabolites by high-performance liquid chromatography and mass spectrometry, *Journal of Agricultural and Food Chemistry*, 2007, 55: 551–560.
17. Antonio, C., Larson, T., Gilday, A., Graham, I., Bergström, E., and Thomas-Oates, J., Hydrophilic interaction chromatography/electrospray mass spectrometry analysis of carbohydrate-related metabolites from *Arabidopsis thaliana* leaf tissue, *Rapid Communications in Mass Spectrometry*, 2008, 22: 1399–1407.
18. Luo, B., Groenke, K., Takors, R., Wandrey, C., and Oldiges, M. J., Simultaneous determination of multiple intracellular metabolites in glycolysis, pentose phosphate pathway and tricarboxylic acid cycle by liquid chromatography-mass spectrometry, *Chromatography A.*, 2007, 1147: 153–164.
19. Plumb, R., Castro-Perez, J., Granger, J., Beattie, I., Joncour, K., and Wright, A., Ultra-performance liquid chromatography coupled to quadrupole-orthogonal time-of-flight mass spectrometry, *Journal of Bioscience and Bioengineering*, 2004, 18: 2331–2337.
20. Soga, T. and Heiger, D. N., Amino acid analysis by capillary electrophoresis electrospray ionization mass spectrometry, *Analytical Chemistry*, 2000, 72: 1236–1241.
21. Soga, T., Ueno, Y., Naraoka, H., Ohashi, Y., Tomita, M., and Nishioka, T., Simultaneous determination of anionic intermediates for *Bacillus subtilis* metabolic pathways by capillary electrophoresis electrospray ionization mass spectrometry, *Analytical Chemistry*, 2002, 74: 2233–2239.
22. Soga, T., Ohashi, Y., Ueno, Y., Naraoka, H., Tomita, M., and Nishioka, T., Quantitative metabolome analysis using capillary electrophoresis mass spectrometry, *Journal of Proteome Research*, 2003, 2: 488–494.
23. Harada, K., Fukusaki, E., and Kobayashi, A., Pressure-assisted capillary electrophoresis mass spectrometry using combination of polarity reversion and electroosmotic flow for metabolomics anion analysis, *Journal of Bioscience and Bioengineering*, 2006, 101: 403–409.
24. Harada, K., Ohyama, Y., Tabushi, T., Kobayashi, A., and Fukusaki, E., Quantitative analysis of anionic metabolites for *Catharanthus roseus* by capillary electrophoresis using sulfonated capillary coupled with electrospray ionization-tandem mass spectrometry, *Journal of Bioscience and Bioengineering*, 2008, 105: 249–260.

25. Soga, T., Igarashi, K., Ito, C., Mizobuchi, K., Zimmermann, H. P., and Tomita, M., Metabolomic profiling of anionic metabolites by capillary electrophoresis mass spectrometry, *Analytical Chemistry*, 2009, 81: 6165–6174.
26. Sugimoto, M., Wong, D. T., Hirayama, A., Soga, T., and Tomita, M., Capillary electrophoresis mass spectrometry-based saliva metabolomics identified oral, breast and pancreatic cancer-specific profiles, *Metabolomics*, 2010, 6: 78–95.
27. Serhan, C. N., Mediator lipidomics, *Prostaglandins and Other Lipid Mediators*, 2005, 77: 4–14.
28. Bamba, T., Shimonishi, N., Matsubara, A., Hirata, H., Nakazawa, Y., Fukusaki, E., and Kobayashi, A., A high throughput and exhaustive analysis of diverse lipids by using supercritical fluid chromatography-mass spectrometry for metabolomics, *Journal of Bioscience and Bioengineering*, 2008, 105: 460–469.
29. Bamba, T., Lee, J. W., Matsubara, A., and Fukusaki, E., Metabolic profiling of hydrophobic compounds lipids by super-critical fluid chromatography/mass spectrometry, *Journal of Chromatography A*, 2012, 1250: 212–219.
30. Yamada, T., Uchikata, T., Sakamoto, S., Yokoi, Y., Nishiumi, S., Yoshida, M., Fukusaki, E., and Bamba, T., Supercritical fluid chromatography/Orbitrap mass spectrometry based lipidomics platform coupled with automated lipid identification software for accurate lipid profiling, *Journal of Chromatography A*, 2013, 1301: 237–242.
31. Uchikata, T., Matsubara, A., Nishiumi, S., Yoshida, M., Fukusaki, E., and Bamba, T., Development of oxidized phosphatidylcholine isomer profiling method using supercritical fluid chromatography/tandem mass spectrometry, *Journal of Chromatography A*, 2012, 1250: 69–75.
32. Matsubara, A., Bamba, T., Ishida, H., Fukusaki, E., and Hirata, K., Highly sensitive and accurate profiling of carotenoids by supercritical fluid chromatography coupled with mass spectrometry, *Journal of Separation Science*, 2009, 32: 1459–1464.
33. Matsubara, A., Harada, K., Hirata, K., Fukusaki, E., and Bamba, T., High-accuracy analysis system for the redox status of coenzyme Q10 by online supercritical fluid extraction-supercritical fluid chromatography/mass spectrometry, *Journal of Chromatography A*, 2012, 1250: 76–79.
34. Becher, J., Muck, A., Mithöfer, A., Svatos, A., and Boland, W., Negative ion mode matrix-assisted laser desorption/ionisation time-of-flight mass spectrometric analysis of oligosaccharides using halide adducts and 9-aminoacridine matrix, *Rapid Communications in Mass Spectrometry*, 2008, 22: 1153–1158.

35. Rosenling, T., Slim, C. L., Christin, C., Coulier, L., Shi, S., Stoop, M. P., Bosman, J., Suits, F., Horvatovich, P. L., Stockhofe-Zurwieden, N., Vreeken, R., Hankemeier, T., and 3 other authors, The effect of preanalytical factors on stability of the proteome and selected metabolites in cerebrospinal fluid (CSF), *Journal of Proteome Research*, 2009, 12: 5511–5522.

36. Sun, G., Yang, K., Zhao, Z., Guan, S., Han, X., and Gross, R. W., Shotgun metabolomics approach for the analysis of negatively charged water-soluble cellular metabolites from mouse heart tissue, *Analytical Chemistry*, 2007, 79: 6629–6640.

37. Vaidyanathan, S., Gaskell, S., and Goodacre, R., Matrix-suppressed laser desorption/ionisation mass spectrometry and its suitability for metabolome analyses, *Rapid Communications in Mass Spectrometry*, 2006, 20: 1192–1198.

38. Miura, D., Fujimura, Y., Tachibana, H., and Wariishi, H., Highly sensitive matrix-assisted laser desorption ionization-mass spectrometry for high-throughput metabolic profiling, *Analytical Chemistry*, 2010, 15: 498–504.

39. Yukihira, D., Miura, D., Saito, K., Takahashi, K., and Wariishi, H., MALDI-MS-based high-throughput metabolite analysis for intracellular metabolic dynamics, *Analytical Chemistry*, 2010, 82: 4278–4282.

40. Miura, D., Fujimura, Y., Yamato, M., Hyodo, F., Utsumi, H., Tachibana, H., and Wariishi, H., Ultrahighly sensitive *in situ* metabolomic imaging for visualizing spatiotemporal metabolic behaviors, *Analytical Chemistry*, 2010, 82: 9789–9796.

41. Giavalisco, P., Hummel, J., Lisec, J., Inostroza, A. C., Catchpole, G., and Willmitzer, L., High-resolution direct infusion-based mass spectrometry in combination with whole ^{13}C metabolome isotope labeling allows unambiguous assignment of chemical sum formulas, *Analytical Chemistry*, 2008, 80: 9417–9425.

42. Weber, R. J., Southam, A. D., Sommer, U., and Viant, M. R., Characterization of isotopic abundance measurements in high resolution FT-ICR and Orbitrap mass spectra for improved confidence of metabolite identification, *Analytical Chemistry*, 2011, 83: 3737–3743.

43. Southam, A. D., Payne, T.G., Cooper, H.J., Arvanitis, T.N., Viant, M.R., Dynamic range and mass accuracy of wide-scan direct infusion nanoelectrospray Fourier transform ion cyclotron resonance mass spectrometry-based metabolomics increased by the spectral stitching method, *Analytical Chemistry*, 2007, 79: 4595–4602.

44. Beckmann, M., Parker, D., Enot, D. P., Duval, E., and Draper, J., High-throughput, nontargeted metabolite fingerprinting using nominal mass flow injection electrospray mass spectrometry, *Nature Protocols*, 2008, 3: 486–504.
45. Roddy, T. P., Horvath, C. R., Stout, S. J., Kenney, K. L., Ho, P. I., Zhang, J. H., Vickers, C., Kaushik, V., Hubbard, B., and Wang, Y. K., Mass spectrometric techniques for label-free high-throughput screening in drug discovery, *Analytical Chemistry*, 2007, 79: 8207–8213.
46. Nanita, S. C., Stry, J. J., Pentz, A. M., McClory, J. P., and May, J. H., Fast extraction and dilution flow injection mass spectrometry method for quantitative chemical residue screening in food, *Journal of Agricultural and Food Chemistry*, 2011, 59: 7557–7568.
47. Fuhrer, T., Heer, D., Begemann, B., and Zamboni, N., High-throughput, accurate mass metabolome profiling of cellular extracts by flow injection-time-of-flight mass spectrometry, *Analytical Chemistry*, 2011, 83: 7074–7080.
48. Boernsen, K. O., Gatzek, S., and Imbert, G., Controlled protein precipitation in combination with chip-based nanospray infusion mass spectrometry: An approach for metabolomics profiling of plasma, *Analytical Chemistry*, 2005, 77: 7255–7264.
49. Schuhmann, K., Almeida, R., Baumert, M., Herzog, R., Bornstein, S. R., and Shevchenko, A., Shotgun lipidomics on a LTQ Orbitrap mass spectrometer by successive stitching between acquisition polarity modes, *Journal of Mass Spectrometry*, 2012, 47: 95–104.
50. Fiehn, O., Wohlgemuth, G., Scholz, M., Kind, T., Lee, d.Y., Lu, Y., Moon, S., and Nikolau, B., Quality control for plant metabolomics: Reporting MSI-compliant studies, *Plant Journal*, 2008, 53: 691–704.
51. Dunn, W. B., Wilson, I. D., Nicholls, A. W., and Broadhurst, D., The importance of experimental design and QC samples in large-scale and MS-driven untargeted metabolomic studies of humans, *Bioanalysis*, 2012, 4: 2249–2264.
52. Sysi-Aho, M., Katajamaa, M., Yetukuri, L., and Oresic, M., Normalization method for metabolomics data using optimal selection of multiple internal standards, *BMC Bioinformatics*, 2007, 8: 93.
53. Bolten, C. J., Kiefer, P., Letisse, F., Portais, J. C., and Wittmann, C., Sampling for metabolome analysis of microorganisms, *Analytical Chemistry*, 2007, 79: 3843–3849.

54. Canelas, A. B., ten Pierick, A., Ras, C., Seifar, R. M., van Dam, J. C., van Gulik, W. M., and Heijnen, J. J., Quantitative evaluation of intracellular metabolite extraction techniques for yeast metabolomics, *Analytical Chemistry*, 2009, 81: 7379–7389.
55. Sugiura, Y., Honda, K., Kajimura, M., and Suematsu, M., Visualization and quantification of cerebral metabolic fluxes of glucose in the awake mice, *Proteomics*, 2013, 14(7–8): 829–838.
56. Gullberg, J., Jonsson, P., Nordstrom, A., Sjostrom, M., and Moritz, T., Design of experiments: An efficient strategy to identify factors influencing extraction and derivatization of *Arabidopsis thaliana* samples in metabolomics studies with gas chromatography/mass spectrometry, *Analytical Biochemistry*, 2004, 331: 283–295.
57. Xia, J. and Wishart, D. S., MSEA: A web-based tool to identify biologically meaningful patterns in quantitative metabolomic data, *Nucleic Acids Research*, 2010, 38: 71–77.
58. Subramanian, A., Tamayo, P., Mootha, V. K., Mukherjee, S., and Ebert, B. L., Gene set enrichment analysis: A knowledge-based approach for interpreting genome-wide, *Proceedings of the National Academy of Science*, 2005, 102: 15545–15550.
59. Xia, J. and Wishart, D. S., Web-based inference of biological patterns, functions and pathways from metabolomic data using MetaboAnalyst, *Nature Protocols*, 2011, 6: 743–760.
60. Chagoyen, M. and Pazos, F., MBRole: Enrichment analysis of metabolomic data, *Bioinformatics*, 2011, 27: 730–731.

3
Sample Preparation

Chapter 3
Sample Preparation

Yoshihiro Izumi, Walter A. Laviña,
and Sastia Prama Putri

Chapter Outline

3.1 Introduction

The accurate measurement of physiological levels of intracellular metabolites is of prime interest in the study of metabolic reaction networks and their regulation *in vivo*. Therefore, it is important to

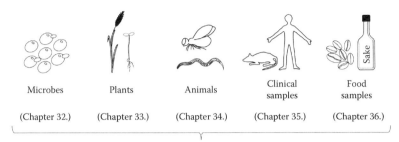

Microbes	Plants	Animals	Clinical samples	Food samples
(Chapter 32.)	(Chapter 33.)	(Chapter 34.)	(Chapter 35.)	(Chapter 36.)

Sample preparation: Harvesting + Quenching + Grinding + Extraction

Figure 3.1 Sample preparation for metabolomics research.

determine the intracellular levels of metabolites accurately. The quality and reliability of metabolomics data will invariably depend on the sampling and sample treatment techniques employed.

There are two important processes in sample preparation: quenching and extraction. Quenching is the process of stopping biological reactions in a cell, and extraction is the process of obtaining metabolites from the cell. Sample quenching has been focused on stopping metabolism at a specific period to measure the true quantity of metabolites at a given time. For quenching, the following are required to measure the quantity of metabolite in a cell accurately: (1) a short timeframe during which the biological reaction is stopped, and (2) limited leakage of metabolite and reproducibility. The appropriate extraction method must be chosen considering the following: (1) cell properties, such as robustness of the cell membrane, (2) chemical properties of the target analyte, and (3) reactivity of enzymes (Putri et al., 2013). In this chapter, we introduce the sample preparation protocol for metabolomics using various specimens namely microbial, plant, animal, medical, and food samples. (See Figure 3.1 and Table 3.1.)

3.2 Microbes

The microbe is an important sample for the field of metabolomics because it has been used for the development of experimental procedures and construction of research tactics

TABLE 3.1
Summary of the Main Quenching Methods for Different Applications

Application	Sample collection	Quenching method	Reference
Microbial cells (*Escherichia coli*)	Membrane filtration	Immersion in methanol used as the extraction solvent	Ohashi et al., 2008
	Membrane filtration (commonly called the "filter culture method")	Immersion in prechilled (−20°C) organic solution used as the extraction solvent	Bennett et al., 2008
	Membrane filtration	Frozen in liquid nitrogen	Unpublished data in our laboratory
Microbial cells (*Saccharomyces cerevisiae*)	Centrifugation (−10°C) after prechilled (−80°C) methanol quenching		Crutchfield et al., 2010
	Membrane filtration	Immersion in prechilled (−20°C) organic solution used as the extraction solvent	Crutchfield et al., 2010
Plant tissues	Membrane filtration	Frozen in liquid nitrogen	Kato et al., 2012
	—	Frozen in liquid nitrogen	Ochi et al., 2012; Yamamoto et al., 2012; Tarachiwin et al., 2008
Animal tissues	—	Frozen in liquid nitrogen	Putluri et al., 2011; Sugimoto et al., 2012

(Putri et al., 2013). Quantitative understanding of microbial metabolism and *in vivo* regulation entails comprehensive coverage of both extracellular and intracellular metabolites. There are two approaches for quantification in metabolomics, namely relative and absolute quantification. Relative quantification normalizes the metabolite signal intensity to that of an internal standard or another metabolite and is typically used in nontargeted large-scale profiling. Absolute quantification uses external standards or internal isotopically labeled standards to determine the absolute metabolite quantity and is mostly used in targeted metabolomics (Lei, Huhman, and Sumner, 2011). In general, the experimental procedures in microbial metabolomics include: (1) biomass cultivation, (2) fast sampling and instant arrest of metabolic activity and deactivation of endogenous enzymatic activity (also widely known as quenching), (3) metabolite extraction, and (4) subsequent quantification of intracellular reactants (metabolites; Mashego et al., 2007; Winder et al., 2008).

Common methods to quantify extracellular metabolites in the cell-free supernatant are either by filtration or centrifugation at low temperatures. In microbial metabolomics, common methods of culturing microbial cells for metabolomic studies include batch liquid culture, continuous culture in a chemostat, and batch culture on a filter support (also known as filter culture). Batch liquid culture is commonly used because of its simplicity and ease, whereas chemostats offer great advantage for controlling culture conditions and reproducibility. The major advantages of filter culture are facile manipulation of the cellular nutrient environment and ease of sampling. Sampling metabolites from either liquid batch or chemostat culture presents similar challenges such as capturing a discrete volume of culture fluid, separating cells from the surrounding medium without altering their metabolome and effectively extracting metabolites from the isolated cells (Crutchfield, Lu, Melamud, and Rabinowitz, 2010). In order to determine the intracellular levels of metabolites accurately, efficient and reliable methods for sample preparation are required. Inasmuch as the key steps involved in sample preparation for metabolome analysis are quenching and extraction of metabolites, this section focuses on several representative procedures for quenching and extraction using the bacterium *Escherichia coli* and yeast *Saccharomyces cerevisiae*.

3.2.1 Working with Bacteria

3.2.1.1 Escherichia coli *as a Model Organism*

The selection of the appropriate quenching and extraction method highly depends on the organism's properties such as size, density, and membrane composition. An ideal quenching procedure should be able to arrest cellular metabolic activity instantly, but at the same time not result in significant cell membrane damage as this may lead to the loss of intracellular metabolites from cells due to leakage. The use of pre-cooled organic solvents such as cold methanol or ethanol has been reported as excellent quenching methods for eukaryotes. However, prokaryotic cells such as *Escherichia coli* tend to leak intracellular metabolites when exposed to the most universally used cold methanol protocol. Furthermore, there was no positive effect on the buffering or ionic strength capability of the quenching solution. (Taymaz-Nikerel et al., 2009). This leakage problem is most probably due to their less robust cell wall and membrane structures as well as composition, which are known to be different from those of eukaryotic cells (Mashego et al., 2007). Therefore, the rapid filtration method is still the preferable choice for prokaryotes quenching even for rapid sampling usage (Bennett, Yuan, Kimball, and Rabinowitz, 2008). Once metabolism has been arrested by quenching, the metabolites need to be extracted from the cells. The goal here is to permanently deactivate endogenous enzymes and permeabilize the cell to release the maximum number of cellular metabolites with the highest possible recoveries (Winder et al., 2008). The most common method is organic solvent extraction, using methanol and chloroform. However, physical stresses, such as high temperature or freeze–thaw, are sometimes applied (Putri et al., 2013). Ideally, the extraction procedure should extract as wide a range of metabolites as possible and the procedure should not modify the intracellular metabolites. As such, physical or chemical modification as well as degradation should be minimized (Winder et al., 2008).

3.2.1.2 Materials

Escherichia coli strains were grown under different cultivation conditions, depending on the study purpose (Ohashi et al., 2008; Bennett et al., 2009).

Sample preparation of *E. coli* for GC/MS and LC/MS analysis ("unpublished data"):

1. Collect the cells for metabolome analysis by transferring the appropriate volume of culture broth to a glass funnel from which the culture medium is removed by vacuum suction through a 47-mm diameter, 0.45-µm pore size PTFE membrane filter (Millipore, MA, USA). Adjust the volume of the culture broth based on OD_{600} at respective sampling points to satisfy the formula: sample volume (mL) $\times$ OD_{600} = 10.

2. Transfer the filter-bound cells into a 2-mL microcentrifuge tube (Eppendorf, Hamburg, Germany), and rapidly cool in liquid nitrogen to quench metabolism.

3. Store the samples at −80°C until extraction.

4. Carry out metabolite extraction from the cells by adding 2 mL of extraction solvent as described in Weckwerth, Wenzel, and Fiehn (2004b; methanol:water:chloroform in 2.5:1:1 v/v/v) with D-camphor-10-sulfonic acid (50 mg mL^{-1}) and adipic acid (8.04 g mL^{-1}) (Wako, Osaka, Japan) added as internal standards.

5. Vortex the samples and incubate for 1 h at −30°C.

6. Remove a portion of the solvent containing extracted metabolites from each tube (350 µL and 700 µL for GC/MS and LC/MS analysis, respectively) and transfer into new tubes.

7. Add distilled water (175 µL for GC/MS and 350 µL for the LC/MS) into the tubes.

8. Centrifuge the mixture at 18,000 $\times$ *g*, 4°C for 5 min to separate the polar and nonpolar phases.

9. Transfer the polar phase hydrophilic metabolites (350 µL for GC/MS and 700 µL for LC/MS) to new 1.5-mL microcentrifuge tubes (Watson, NJ, USA).

10. For LC/MS analysis, filter the polar phase with 0.2-µm pore size PTFE filter (Millipore, MA, USA) before transferring to new 1.5-mL tubes.

11. Remove the methanol from the samples by centrifugal concentration for 2 h using a VC-96R Spin Dryer Standard (Taitec, Tokyo, Japan) followed by overnight lyophilization in a VD-800F Freeze Dryer (Taitec, Tokyo, Japan).

12. Store the lyophilized samples at −80°C until analysis.

13. For GC/MS analysis, derivatize the extracted metabolites by oximation and silylation prior to analysis. First, dissolve the oximation reagent, methoxyamine hydrochloride (Sigma-Aldrich, MO, USA) in pyridine (Wako, Osaka, Japan) to a concentration of 20 mg mL^{-1} and add 50 µL to each sample tube containing the lyophilized extracts.

14. After reaction at 30°C, 1,200 rpm for 90 min, add 25 µL of N-methyl-N-(trimethylsilyl)trifluoroacetamide (MSTFA; GL Sciences, Tokyo, Japan) and incubate at 37°C, 1,200 rpm for 30 min. Transfer the derivatized samples to glass vials (Chromacol, Hertfordshire, UK) and analyze within 24 h.

15. For LC/MS analysis, dissolve the extracted metabolites in 30 µL of ultrapure water, vortex, and transfer to glass vials.

Sample preparation of *E. coli* for capillary electrophoresis mass spectrometry (CE/MS) analysis (Ohashi et al., 2008):

1. Filter the culture with approximately 10^9 cells (calculated as optical density at 600 nm × sampling volume of culture (mL) = 20) by a vacuum filtration system using a 0.4-mm pore size filter.

2. Wash the residual cells on the filter twice with 5 ml of Milli-Q water.

3. Immerse the filter in 2 mL of methanol including 5 mM each of internal standards, methionine sulfone, and D-camphor-10-sulfonic acid.

4. Sonicate the dish for 30 s using an Elma Transsonic T460/H ultrasonic syringe (not an ultrasonic cell disrupter) (Elma Hans Schmidbauer GmbH & Co., Singen, Germany) to suspend the cells completely.
5. Transfer a portion of the methanol cell suspension (1.6 mL) to a Falcon Blue Max Jr., 352097 centrifugal tube (15 mL; Becton Dickinson & Co., NJ, USA), and mix with 1.6 mL of chloroform and 640 mL of Milli-Q water.
6. After vortexing well, centrifuge the mixture at 4,600 × g and 4°C for 5 min. Distribute the aqueous layer (750 mL) to three Amicon Ultrafree-MC ultrafilter tips (Millipore Co., MA, USA) and centrifuge at 9,100 × g at 4°C for approximately 2 h.
7. Dry the filtrate and preserve at −80°C until CE/MS analysis.
8. Prior to analysis, dissolve the sample in 25 ml of Milli-Q water.

Note:: This protocol can be optimized by introducing a sonication step prior to extraction to improve quantitative analyses of phosphate-rich metabolites and other hydrophilic charged metabolites.

Sample preparation of *E. coli* for absolute quantitation: filter culture (Bennett et al., 2008, 2009; Rabinowitz and Kimball, 2007; Yuan et al., 2009):

1. Grow cells in a medium containing nutrients labeled with stable isotopes. Perform quenching and metabolite extraction in solution spiked with unlabeled standards.
2. Calculate the concentration of metabolites in the cells using the ratio of the labeled extracted metabolite to the unlabeled internal standard.
3. To enable rapid noninvasive alteration of the extracellular nutrient environment, use the filter culture method in which *E. coli* are grown on nitrocellulose filters on top of the agarose plates (Figure 3.1. Overview of workflow, illustrated for the case of nonadherent cells such as *E. coli*; Bennett et al., 2008).

4. To establish the filter cultures, filter the exponentially growing *E. coli* (5-mL/filter, OD_{650} ~ 0.1) onto 82-mm, 0.45-µm pore size nylon membrane (MAGNA, GE Osmonics, Minnetonka, MN).

5. Place the filters face up on agarose plates loaded with complete minimal media.

6. To prepare the plates, wash the ultra pure agarose three times in cartridge-purified water to remove trace contaminants and add to the desired liquid media at 1.5% by weight.

7. Separate the cells from growth media and perform quenching by placing them cell side down in a 100-mm polystyrene tissue culture dish filled with 2.5 mL of cold organic solvent (a mixture of acetonitrile:methanol:water (2:2:1, v/v/v) with 0.1% formic acid, prechilled –20°C).

8. After 15 m, remove the cell–solvent mixture from the dish and set aside.

9. Use an additional 1 mL of solvent to wash the filter to ensure that all the cellular material has been removed, and combine the resulting solution with the initial 2-mL extract.

10. Split the resulting 3-mL volume into two Eppendorf tubes and centrifuge for 5 min at maximum speed and 4°C, then set aside the soluble extract.

11. Re-extract each of the residual pellets twice with 50 µL of solvent at 4°C for a total extraction volume (after absorptive solvent losses) of 3 mL.

12. For the water-immiscible solvents, the extraction procedure is the same as above, except that the two phases are divided after conclusion of the extraction steps.

13. Analyze the phases separately by LC/MS/MS. For the chloroform:methanol mixture, the upper (aqueous) phase is 200 µL of the 700-µL total volume.

14. For ethyl acetate, the lower (aqueous) phase is 100 µL of the 700-µL total volume.

15. For almost all of the studied (generally hydrophilic) compounds, the vast majority of the signal is associated with the aqueous phase.

16. Once an extract is obtained, it can be analyzed using a wide variety of chromatography-MS procedures. The authors recommend the use of LC/MS/MS because LC has broad diversity of analytes without the need for derivatization. Furthermore, a key advantage of electrospray ionization (ESI) is its efficiency in converting charged compounds into gas phase ions.

Note:

1. This method also works well with other microorganisms, such as *S. cerevisiae*.
2. The solvent mixture of acetonitrile:methanol:water (2:2:1, v/v/v) with 0.1% formic acid was selected to minimize degradation of high-energy metabolites such as ATP and NADH during the quenching and extraction process.
3. Other extraction protocols such as the methanol extraction method is recommended for extracting amino acids in *E. coli* (100% methanol for the first extraction and methanol:water (4:1. v/v) for the subsequent two extractions).

Chapter 3.2.2 Working with Yeast

3.2.2.1 Saccharomyces cerevisiae *as a Model Organism*

Saccharomyces cerevisiae is a model organism for eukaryotes and is one of the most important hosts for the production of biomaterials, including fine chemicals and biofuels (Kato, Izumi, Hasunuma, Matsuda, and Kondo, 2012). Given the importance of yeast as a research sample, there is myriad research on the optimization of quenching and extraction methods for yeast cells (Kawase, Tsugawa, Bamba, and Fukusaki, 2014; Kato et al., 2012; Villas-Bôas et al., 2005). There are two commonly used methods for quenching in yeast: cold methanol and fast filtration. In the standard approach, cold methanol quenching involves mixing culture media directly with cold ($\leq$–40°C) methanol to quench metabolism, centrifuging in a prechilled rotor ($\leq$–20°C) to isolate the cells, and subsequently extracting the cell pellet. This method was first described by de Koning and van Dam (de Koning and van Dam, 1992). In this method, quenching is achieved by initially mixing the cells with cold methanol resulting in a cold-induced reduction of reaction rates and/or denaturation of organic-induced enzyme. Nevertheless, it aims to avoid the following: the effect of cold-induced ice crystal formation that punctures membranes, and organic-induced membrane dissolution of the cells, which ultimately causes membrane disruption and leakage of metabolites from the cells. However, further exposure to organic solvent after quenching cells results in membrane disruption and extraction of metabolites from the cells. The main risk in this method is the loss of metabolites (e.g., due to cell leakage) during the centrifugation step. Another disadvantage is the somewhat laborious nature of the process. An alternative approach is fast filtration, which involves the separation of the cells from medium prior to quenching metabolism. The cell-loaded filter is then simultaneously quenched and extracted by placing it into cold extraction solvent. Due to the lag time during filtration, the main downside of this approach is the potential for metabolome changes during filtration (Crutchfield et al., 2010; Villas-Bôas et al., 2005).

3.2.2.2 Materials

Saccharomyces cerevisiae strains are grown under different cultivation conditions, depending on the study purpose. Please refer to the references for details on cultivation conditions.

Sample preparation of *S. cerevisiae*: harvesting cells by centrifugation after methanol quenching (Crutchfield et al., 2010):

1. Prechill a centrifuge rotor capable of handling 50-mL centrifuge tubes to −80°C.
2. Directly quench 10-mL culture broth into 20-mL −80°C methanol in a 50-mL centrifuge tube.
3. Spin down at ~ 2,000 × g in centrifuge cooled to −10°C for 5 min.
4. Prepare extraction solution: acetonitrile: methanol: water (2:2:1. v/v/v), or alternatively methanol:water (4:1. v/v) (both give roughly equivalent results for

methanol-quenched cells). All solvents should be of the highest purity available (minimum HPLC grade).

5. Cool extraction solution to −20°C. Shake by mixing prior to use.

6. When ready to extract (i.e., after pouring off supernatant following the initial quenching step), pipette 400-µL of cold extraction solvent (−20°C) directly onto pellet.

7. After pipetting up and down (do not vortex), transfer mixture to a 1.5-mL Eppendorf tube and let it sit on ice for 15 min.

8. Spin down the mixture in a microcentrifuge (highest speed, ~16,000 × *g*) for 5 min to pellet cell debris.

9. Transfer the supernatant to a separate 1.5-mL Eppendorf tube, recording the volume recovered.

10. Resuspend the pellet in a volume of extraction solvent equal to the difference of the recovered supernatant and 800 µL. The pellet is typically difficult to resuspend. Prefilling the pipette tip with the 100 µL of extraction solvent and gently perturbing the pellet with the pipette tip before depressing to release the extraction solvent can aid in resuspension. Take care not to clog the pipette tip with sticky cell debris.

11. Keep the resuspended mixture on ice for an additional 15 m.

12. Spin down sample and pool the supernatant with the prior fraction.

13. Vortex pooled mixture and analyze.

Sample preparation of *S. cerevisiae*: harvesting cells by vacuum filtration method (Crutchfield et al., 2010):

1. Construct a filtration apparatus as follows. Seal the top of a 15-mL centrifuge tube with a two-hole rubber stopper for collection of the culture medium. Connect one hole to the vacuum and the other to the filter support. The filter should sit on top of the glass (or metal) frit, with the open-bottom graduated cylinder attached by a clamp. The filter must cover the entire frit, or otherwise cells will be lost during filtration and quantitation will be unreliable. To initiate filtration, pour the cells

into the graduated cylinder at the top of the apparatus. Once filtration is complete, remove the clamp and quickly transfer the filter to the quenching solvent.

2. Thoroughly rinse the apparatus with purified water.

3. Place a 25-mm 0.45-µm pore size nylon filter on the filter base and prewet with purified water.

4. Connect a 15-mL centrifuge tube to the rubber stopper at the bottom of the filtration apparatus, forming a tight seal. This tube will be used to collect the extracellular media.

5. Measure 10 mL of culture, using either a 15-mL centrifuge tube or a volumetric pipette. The recommended culture density at time of extraction is ~4 × 10^7 cells/mL (Klett ~ 130 or OD_{650} ~ 0.5).

6. Pour the culture immediately into the glass cylinder at the top of the filtration apparatus. Filtration should occur rapidly.

7. When filtration appears complete, wait ~1 s and then remove the clamp to free the filter.

8. Immediately place the filter, cells side down, into a 35-mm Petri dish containing 700 µL of prechilled (–20°C) extraction solvent (acetonitrile:methanol:water (2:2:1, v/v/v)). Time from initiation of sampling to quenching should not exceed 30 s.

9. Allow extraction to proceed (stirring not required) at –20°C for 15 min.

10. Collect as much solvent, cells, and debris as possible into a 1.5-mL Eppendorf tube. To remove cells and debris from the filter, flip it cell side up and wash 10 times with the pooled solvent at the bottom of the Petri dish. To release adherent solvent from the filter, dab it on a dry part of the Petri dish approximately five times. From this point forward the sample can be kept on ice.

11. Spin down the mixture in a microcentrifuge (highest speed, ~16,000 × g) for 5 min to pellet cell debris.

12. Transfer the supernatant to a separate 1.5-mL Eppendorf tube and resuspend the remaining pellet in 100 µL fresh extraction solvent.

Note: The pellet is typically difficult to resuspend. Prefilling the pipette tip with 100 μL of extraction solvent and gently perturbing the pellet with the pipette tip before depressing to release the extraction solvent can aid resuspension. Take care not to clog the pipette tip with sticky cell debris.

1. Keep the resuspended mixture on ice for an additional 15 m.
2. Spin down sample and pool the supernatant with the prior fraction.
3. Vortex pooled mixture and analyze.

Sample preparation of *S. cerevisiae*: harvesting cells by vacuum filtration method (Kato et al., 2012):

1. Filter the fermentation liquid (3 mL) containing 100–150 mg (fresh weight) of yeast cells through a membrane filter (Omnipore, 0.45-μm pore size, 25-mm diameter polytetrafluoroethylene, Millipore, MA).
2. Immediately after filtration, wash the yeast cells with 5 mL of cold aqueous ammonium hydrogen carbonate (50 mM).
3. Transfer the filter into an Eppendorf tube with a membrane filter, and freeze in liquid nitrogen.
4. Extract 10 mg of lyophilized cells with 0.9 mL of methanol:chloroform:water (2.5:1:1, v/v/v) using a Shake Master NEO (Bio Medical Science, Tokyo, Japan) with a zirconia bead for 5 min at 1,500 rpm.
5. Incubate the mixture continuously in a Thermomixer comfort (Eppendorf, Hamburg, Germany) at 15°C and 1,200 rpm for 30 min, followed by centrifugation at 4°C and 16,000 × g for 3 min.
6. Mix the supernatant (630 μL) containing hydrophilic metabolites with 280 μL of Milli-Q water, and recentrifuge under the same conditions.
7. Dry two aliquots (300 μL) of the aqueous layer under vacuum and store at −80°C until GC/MS and LC/MS analysis.

Sample preparation of *S. cerevisiae* for single-cell metabolite analysis using the microarrays for mass spectrometry (MAMS) platform by MALDI-MS (Ibáñez et al., 2013):

1. Harvest the cells of the main culture grown at an OD_{600} of 1.0 to 1.2.
2. Perform quenching by injecting 1 mL of yeast culture in 4 mL of a methanol:water (3:2, v/v) with 0.85% (w/v) ammonium bicarbonate buffer, at −40°C.
3. Rapidly submerge the Falcon tubes containing the aliquot in a −40°C glycerol bath for 1 min to ensure quenching of the metabolism.
4. Centrifuge the samples for 1 min at −10°C at 1,000 × g in a nonfixed rotor centrifuge.
5. Remove the supernatant (growth medium), and shock freeze the Falcon tube containing the cell pellet in liquid nitrogen and store at −80°C until the measurement.
6. Prior to spotting the cells on the MAMS target for MALDI-MS measurement, reconstitute the cells in 500 µL of a methanol:water (3:2, v/v) plus 0.85% (w/v) ammonium bicarbonate buffer, precooled at −40°C, and carefully shake to avoid cell aggregation.
7. Transfer the samples to precooled Eppendorf vials (2 mL).
8. Centrifuge the Eppendorf vials at −10°C for 1 m at 1,000 × g (fixed rotor).
9. Remove the supernatant to avoid any remaining growth medium trapped in the cell pellet.
10. Reconstitute the cell pellet in approximately 50 µL of a methanol:water (3:2, v/v) (−40°C) and slightly shake.
11. Spread two microliters of the cell suspension (approx. 2.5×10^7 cells per mL) over a cooled MAMS plate (−4°C).
12. Further details on MALDI-MS measurement can be seen in Ibáñez et al. (2013).

3.3 Plants

Plants display enormous diversity in their metabolism. This is because approximately 50,000 different compounds have been elucidated in plants (Aharoni et al., 2000), and it is predicted that the final figure for the plant kingdom will approach or even exceed 200,000 (Fiehn, 2002). Therefore, plant metabolomics represents a considerable challenge for scientists. The application of metabolomics techniques to plant science was pioneered by the group at the Max Planck Institute. Weckwerth and coworkers (2004) demonstrated that plant metabolomics can be an invaluable diagnostic tool for biochemical phenotyping of biological systems. Plant metabolomics technique has already been used as a powerful tool for functional genomics, precise phenotyping, and industrial applications.

This section introduces the representative sample preparation protocols for the model plant, *Arabidopsis thaliana*, and practical plants, such as *Glycine max* and *Angelica acutiloba*.

3.3.1 Working with Roots of Plants

3.3.1.1 Angelica acutiloba *(Traditional Herbal Medicine)*

Traditional Chinese medicine (TCM) has been widely used for several centuries in Asian countries including China, Korea, and Japan because of its therapeutic effects. The roots of *Angelica acutiloba* (*yamato-toki* in Japanese) have traditionally been used in the treatment of gynecological diseases such as menoxenia, arthritis, and anemia because of their hematopoietic, analgesic, and sedative effects (Haruki, Hiroaki, Jong-Chol, and Yasuo, 1985). For these reasons, *yamato-toki* roots are widely commercialized in Japan. However, these *Angelica* roots exhibit variations in their chemical constituents and pharmacological effects based on the differences in species, geographical conditions, as well as processing methods (Piao, Park, Cui, Kim, and Yoo, 2007). In addition, as *yamato-toki* roots are traditional medicines sold primarily by professional herbalists, their commercial values are determined by subjective observations of smell, taste, and appearance (Tarachiwin, Katoh, Ute, and Fukusaki, 2008). This makes quality assessment in the market extremely difficult and impractical when considering mass production of this product. Therefore, the need for a practical and standardized quality assessment method for industrialized *yamato-toki* roots is a critical topic (Tianniam, Tarachiwin, Bamba, Kobayashi, and Fukusaki, 2008).

A combination of metabolic fingerprinting technique and multivariate analysis can be used instead of specific marker compounds to assess the quality corresponding to the cultivation area, which gives faster and more reliable results compared to those obtained from the sensory test (Tarachiwin et al., 2008; Tianniam, Bamba, and Fukusaki, 2010; Tianniam et al., 2008).

3.3.1.2 Materials

Dried root samples of *A. acutiloba* (*yamato-toki*) are provided by a folk medicine company (in this case, Fukuda Shoten, Nara, Japan). The sensory qualities of the samples are judged by a professional taster who has vast expertise and experience in sensory evaluation (in this case, from the Fukuda Shoten company).

The toki roots are ground into powder form with a Wonder Blender (Osaka Chemical, Osaka).

The ground samples are vacuum sealed and kept under −30°C until the sample preparation process.

Sample preparation of *Angelica* roots for ¹H NMR analysis (Tarachiwin et al., 2008):

1. Add 1.5 mL of deuterium oxide (D_2O, D 99.9% atom%) to 150 mg of dried toki roots in a 2-mL Eppendorf tube.
2. Incubate the mixture continuously and shake in a Thermomixer comfort (Eppendorf, Hamburg, Germany) at 70°C, 1,400 rpm for 3 h, followed by centrifugation at 25°C and 20,000 × *g* for 30 min.
3. Transfer the supernatant (300 µL) containing hydrophilic metabolites to a new tube and mix with 300 µL of 0.2-M phosphate buffer solution containing 3-mM 3-(trimethylsilyl)-1-propanesulfonic acid solution (internal standard) to give a 600-µL solution for NMR measurement.
4. All samples are prepared in 1 day and stored at 4°C prior to the analysis.

Sample preparation of Angelica roots for hydrophilic metabolites profiling with GC/MS (Tianniam et al., 2008):

1. To the ground sample (40 mg), add 1 mL of methanol:chloroform:water (2.5:1:1, v/v/v) and 60 µL of ribitol (20 mg mL^{-1}) as internal standard.
2. Homogenize the mixture and disrupt using the MM 301 mixer mill (Retsch GmbH, Haan, Germany) at 20 Hz for 5 min, and centrifuge at 15,000 × g under 4°C for 3 min.
3. Transfer the supernatant (900 µL) containing hydrophilic metabolites into a new tube and mix with 400 µL of Milli-Q water, then recentrifuge under the same conditions.
4. Transfer the aqueous layer (400 µL) to a clean tube, and evaporate the aqueous layer extracts under vacuum using a centrifugal concentrator at room temperature for approximately 2 h, followed by a drying process in a freeze-dryer overnight before derivatization.
5. For oximation, add 50 µL of methoxyamine hydrochloride in pyridine (20 mg mL^{-1}) to the dried hydrophilic crude extract, and incubate in Thermomixer comfort (Eppendorf) at 1,200 rpm for 90 min at 30°C.
6. Add 100 µL of N-methyl-N-(trimethylsilyl)triluoroacetamide (MSTFA) and incubate at 37°C for 30 min in order to induce the silylation reaction before injecting to the GC/MS machine.

3.3.2 Working with Leaf of Plants

3.3.2.1 Glycine Max

Soybean (*Glycine max*) is an important crop for human and domestic animal nutrition due to its high protein and lipid content. It is estimated that about 69% of the world's dietary proteins and 30% of the world's edible oils are from soybean (Luo, Yu, and Liu, 2005). Soybean is also widely used for production of many kinds of processed foods popular in Japan, such as tofu, natto, edamame, and nimame (Lee et al., 2012). For the food industry, it is important to understand the characters and adequate usages of soybean cultivars. Therefore, a metabolomics approach is useful for obtaining the detailed phenotype of soybean cultivars.

On the other hand, it is well known that environmental alterations such as salt stress could reduce plant height and leaf size, inhibit nitrogen fixation, and decrease protein content and seed quality, thus causing significant reduction in growth and yield of soybean. Lu et al. (2013) demonstrated that MS-based metabolomics can provide a fast and powerful approach to discriminate the salt-tolerance characteristics of soybeans. Therefore, advances in metabolic profiling would open up the possibility of gaining previously unobtainable insights into the physiological adaptations of plants to environmental alterations.

Sample preparation of soybean leaf for GC/MS and LC/MS analyses (Lu et al., 2013):

3.3.2.2 Materials

Soybeans were cultured in 1 × Hoagland nutrient solution in a greenhouse or growth chambers with light condition.

1. Harvest fresh soybean leaves and immediately freeze in liquid nitrogen.
2. Perform homogenization using mortar and pestle in liquid nitrogen, after which 150 mg of pooled homogenized plant material is weighed in an Eppendorf tube.
3. To the ground materials, add 300 µL of cold methanol: water (4:1, v/v) containing 1.1 mM ribitol as internal standard.
4. Incubate the mixture in a Thermomixer comfort (Eppendorf, Hamburg, Germany) at 1,200 rpm for 15 min at 4°C.
5. Sonicate the extract for 5 min and centrifuge at 20,000 × g under 4°C for 15 min.
6. A portion of the supernatant (300 µL) was transferred to a new tube and used for LC/MS analysis.
7. For GC/MS analysis, a 50-µL aliquot of the supernatant was further derivatized by methoxyamination and trimethylsilylation using methoxyamine hydrochloride in pyridine (20 mg mL^{-1}) and MSTFA, respectively.

Sample preparation of soybean for lipid profiling (Lee et al., 2012):

1. Grind 12 soybeans using a high-speed vibrating sample mill (TI-100, CMT Co. Ltd., Tokyo, Japan) at 50 Hz for 1.5 min at room temperature.
2. Immerse the soybean powder (20 mg) in liquid nitrogen and grind again for 1 min at 20 Hz in a ball mill mixer MM301 (Retsch, Haan, Germany).
3. Extract soybean lipids with 500 µL of chloroform:methanol (2:1, v/v) with mixing, then sonicate for 3 min, and centrifuge at $10,000 \times g$ for 5 min.
4. Subject the supernatant to supercritical fluid chromatography coupled with mass spectrometry analysis.

3.3.3 Working with Shoots and Roots of Plants

3.3.3.1 Arabidopsis thaliana

Arabidopsis thaliana is a small flowering plant that is widely used as a model organism in plant biology. Although not of major agronomic significance, *Arabidopsis* offers important advantages for basic research in genetics and molecular biology (Somerville and Koornneef, 2002), namely: (1) the genetic and genomic methods and resources (mutant lines) are available (Meinke, Cherry, Dean, Rounsley, and Koornneef, 1998), (2) the small size and simple growth requirements of *Arabidopsis* make it easy to grow in laboratory conditions, (3) the lifecycle is short, about eight weeks from germination to seed maturation, (4) transformation is efficient by utilizing *Agrobacterium tumefaciens*. The metabolomics approach has been successfully performed in *A. thaliana* for functional genomics studies on complex metabolic networks, their modes of action and their role in metabolism.

Fiehn et al. (2000a) discovered 326 metabolites in *A. thaliana* leaf extracts (101 polar and 63 lipophilic identified compounds, plus 113 polar and 49 lipophilic compounds of unknown chemical structure) using GC/MS, and demonstrated the use of metabolite profiling as a tool to extend and enhance the power of existing functional genomics approaches significantly.

Although recent progress in phytochemical genomics studies using *Arabidopsis* has enriched the list of functionally identified genes with the aid of transcriptome resources (Hirai et al., 2007), the majority of metabolic gene functions as well as plant metabolic systems themselves remain unknown, largely because the phytochemicals produced in *Arabidopsis* have not been fully characterized. Matsuda et al. developed a spectral tag (MS2T) library-based peak annotation procedure for *Arabidopsis* nontargeted secondary metabolites profiling analysis using LC/MS (Matsuda et al., 2009). Furthermore, the integrated analysis of gene expression and secondary metabolites in Arabidopsis was intensively performed for construction of the AtMetExpress development data set, which is part of the AtMetExpress metabolite accumulation atlas (Matsuda et al., 2010). This study found that the functional differentiation of secondary metabolite biosynthesis among the various tissues was achieved by controlling the expression of a small number of key regulatory genes. It also postulated that a simple mode of regulation called transcript-to-metabolite is the origin of the dynamics and diversity of plant secondary metabolism.

In the near future, systems biology, metabolomics-driven data, and other omics will play key roles in understanding plant systems and developing further biotechnology applications.

3.3.3.2 *Materials*

In our case, *Arabidopsis* ecotype Columbia and T-DNA-inserted knockout mutants were obtained from the Arabidopsis Biological Resource Center, ABRC (Ohio State University, USA; Rhee et al., 2003).

Seedlings of *Arabidopsis* strains are grown in pots containing soil at 20°C with a 16 h daily photoperiod.
Sample tissues are weighed and stored at −80°C until extraction of metabolites.

Sample preparation of *Arabidopsis* leaves for metabolic profiling with GC/MS (Fiehn et al., 2000a,b):

1. Weigh fresh leaves (300 mg) in a 2-mL Eppendorf tube and immediately freeze in liquid nitrogen for quenching.

2. Homogenize the plant materials by disrupting with MM 301 mixer mills (Retsch GmbH, Haan, Germany).
3. Add 1.4 mL of methanol, 50 µL of water, and internal standards including $^{13}C_{12}$-sucrose, $^{13}C_6$-glucose, d_8-glycerol, d_4-ethanolamine, d_6-ethylene glycol, d_3-aspartate, $^{13}C_5$-glutamate, d_4-alanine, d_8-valine, d_3-leucine, and d_5-benzoic acid.
4. Carry out extraction at 70°C for 15 min, and centrifuge at $14,000 \times g$ for 3 min.
5. Decant the supernatant into a screw-top glass tube; then add 1.4 mL of water and 0.75 mL of chloroform.
6. Vortex the mixture, and centrifuge for 10 min at $1,400 \times g$.
7. Dry the methanol/water phase in a SpeedVac concentrator overnight.
8. To protect carbonyl moieties by methoximation or ethoximation, add 50 µL of alkoxyamine hydrochloride in pyridine (20 mg mL^{-1} solution) and incubate at 30°C for 90 min.
9. Derivatize acidic protons by adding 50 µL of N-methyl-N-tert-butyldimethylsilyltrifluoroacetamide (MTBSTFA; 70°C for 30 min.) or N-methyl-N-trimethylsilyltrifluoroacetamide (MSTFA) at 37°C for 30 min.

Sample preparation of *Arabidopsis* seedlings for metabolic profiling with GC/MS (Jumtee et al., 2008):

1. Weigh fresh leaves (300 mg) in a 2-mL Eppendorf tube and immediately freeze in liquid nitrogen for quenching.
2. Homogenize the plant materials by disrupting with MM 301 mixer mills (Retsch GmbH, Haan, Germany).
3. Extract the metabolites using 1 mL of methanol: water: chloroform (2.5:1:1, v/v/v) and 40 µL of ribitol (200 mg mL^{-1}) as internal standard. The extraction process was performed with a MM 301 mixer mill (Retsch GmbH, Haan, Germany) at 20 Hz for 5 min, followed by centrifugation at $16,000 \times g$ under 4°C for 10 min.
4. Transfer the supernatant (800 µL) containing hydrophilic metabolites to a new Eppendorf tube and mix with 400 µL of Milli-Q water, followed by centrifugation under the same conditions.

5. Transfer the aqueous layer (900 μL) to a new Eppendorf tube, and evaporate the aqueous layer extract under vacuum using a centrifugal concentrator at room temperature for approximately 2 h, followed by a drying process in a freeze-dryer overnight before derivatization.

6. For oximation, add 100 μL of methoxyamine hydrochloride in pyridine (20 mg mL^{-1}) to the dried hydrophilic crude extract, and incubate in a Thermomixer comfort (Eppendorf) at 1,200 rpm for 90 min at 30°C.

7. Add 100 μL of N-methyl-N-(trimethylsilyl)triluoroacetamide (MSTFA) and incubate again at 37°C for 30 min in order to induce the silylation reaction before injecting to the GC/MS machine.

Sample preparation of *Arabidopsis* tissues for secondary metabolite profiling with LC/MS (Matsuda et al., 2010, 2009):

1. Homogenize the frozen tissues in five volumes of 80% aqueous methanol containing 0.5 mg mL^{-1} lidocaine and D-camphor sulfonic acid (internal standard) using a MM 300 mixer mills (Retsch GmbH, Haan, Germany) with a zirconia bead for 6 min at 20 Hz.

2. Centrifuge at $15,000 \times g$ for 10 min and filter the supernatant using an Ultrafree-MC filter (0.2-μm pore size; Millipore, Billerica, MA, USA).

3.4 Animals

Metabolomics technology has been applied to study biological phenomena in several model animals, including fruit fly, nematode, and zebra fish. Comprehensive analysis of metabolites in these organisms has provided a wealth of information on physiological, developmental, and pathological processes. Metabolomics can therefore provide novel insights that can be developed and applied to research in other species. This section introduces the representative sample preparation protocols for model animal metabolomics.

3.4.1 *Working with* Drosophila melanogaster

Although human metabolomics is necessarily observational, studies of simpler organisms offer the prospect of reconciling experimental genetic lesions with their impact on the metabolome, both in order to understand existing pathways and to elucidate new metabolic networks. Of the metazoans, the fruit fly (*Drosophila melanogaster*) offers perhaps the best balance between genetic tractability, availability of well-characterized genetic mutant stocks, and organismal complexity (Kamleh, Hobani, Dow, and Watson, 2008). In addition, Drosophila is readily cultured under a controlled environment and provides a method for the discernment of genuine metabolic differences, rather than chance differences relating to more complex environmental factors (Malmendal et al., 2006). An example of a metabolomics-based approach in *D. melanogaster* is the study on the effects of inbreeding on the *D. melanogaster* metabolome; perturbations or some mutations in metabolic pathways have been studied in order to investigate the relationship between metabolite and phenotype. Pedersen et al. (2008) applied metabolomics using NMR to study the effects of inbreeding and temperature stress on the *Drosophila* metabolism. They found that the metabolite fingerprints of inbred and outbred flies were clearly distinguishable. The metabolomic effect of inbreeding at benign temperature was related to gene expression data from the same inbred and outbred lines. Both gene expression and metabolite data indicate that fundamental metabolic processes are changed or modified by inbreeding.

Watson et al. have performed LC/MS-based metabolomics for *D. melanogaster* functional genomics where they have looked at the impact of the rosy (*ry*) (Kamleh et al., 2008), chocolate (*cho*) (Kamleh, Hobani, Dow, Zheng, and Watson, 2009), and yellow (*y*) (Al Bratty, Chintapalli, Dow, Zhang, and Watson, 2012) mutations on the Drosophila metabolome. The observation of a build-up of metabolites upstream of the genetic lesion is strongly reminiscent of the mutational systems biology approach. Thus, completely unexpected impacts of gene mutation were identified, showing that, even for an exhaustively studied locus, metabolomic analysis can provide new insights.

3.4.1.1 Materials

Fly lines used in this protocol were obtained from the Bloomington Drosophila Stock Center, BDSC (Indiana University, USA; Cook, Parks, Jacobus, Kaufman, and Matthews, 2010).

- The flies are held under well-controlled laboratory conditions (25°C and 12:12-h light–dark cycle) on a standard agar-sugar-yeast-oatmeal *Drosophila* medium (Bubliy and Loeschcke, 2005).
- Every subsequent generation is founded by a mix of parents collected from different bottles. In this example, there are 25 bottles in total with about 50 pairs of parents per bottle.
- The experimental lines are established by flies from the fourth generation of the mass population.
- After flies are anesthetized by chilling on ice, the flies are transferred to microcentrifuge tubes and snap-frozen in liquid nitrogen for quenching of the metabolism. Samples are stored at −80°C until extraction of metabolites.

Sample preparation of *Drosophila* for ¹H NMR analysis (Malmendal et al., 2006; Pedersen et al., 2008):

1. Homogenize the frozen samples (50 flies) in 400 µL ice-cold acetonitrile (50%) and centrifuge at 17,000 × g for 10 min at 5°C.
2. Wash a 10-kDa microfilter (Ultrafree-MC; Millipore, Billerica, MA, USA) four times with distilled water before use to remove glycerol (used for preservation of the microfilters). Pass the supernatant through the 10-kDa microfilter.
3. Spin the samples for 45 min at 6,000 × g, and lyophilize the filtered supernatant before storing at −80°C until NMR analysis.
4. Immediately before the NMR measurements, rehydrate the samples in 650 µL of 50 mM phosphate buffer made up in D_2O (pH 7.4), and transfer 600 µL to a 5-mm NMR tube. The buffer should contain 50 mg mL⁻¹ of the chemical shift reference dimethylsilapentanesulfonic acid.

Sample preparation of *Drosophila* for LC/MS measurement (Chintapalli, Al Bratty, Korzekwa, Watson, and Dow, 2013; Kamleh et al., 2008, 2009):

1. Add ice cold methanol:water:chloroform (3:1:1, v/v/v) solvent at 0°C (250 μL) to the frozen samples (10 flies).
2. Homogenize the flies for 30 s by using an ultrasonic cell disruptor (Misonix, Inc., NY, USA)
3. Centrifuge the homogenates for 10 min at 4°C.
4. Remove the extract from the cell debris and store at −80°C until LC/MS analysis.

3.4.2 *Working with* Caenorhabditis elegans

The nematode *Caenorhabditis elegans* (commonly called "worm") is one of the best-studied animals in science primarily because of the range of relatively simple experimental protocols that allow extremely detailed manipulations. The animals grow easily on agar plates or liquid culture with *Escherichia coli* as its food, and under laboratory conditions have a generation time of 3.5 days from fertilized egg to reproducing adult. The *C. elegans* has also been characterized in great detail in terms of its development, lifespan, morphology, and physiology at the cellular level. Genetics are especially well developed in *C. elegans*, which has both self-fertilizing hermaphrodites and males and thus allows great flexibility in establishing and maintaining novel genetic lines. In addition, *C. elegans* is one of the few animals for which high-throughput *in vivo* RNA interference (RNAi) screening has been established (Kamath et al., 2003). *C. elegans* is a good platform, for example, to understand the physiology of lifespan and aging, for the study of genetic diseases, drug toxicity screening, and pharmacological studies.

Blaise et al. (2007) have examined metabolomic differences between mutant *C. elegans* where they used ^{1}H high-resolution magic angle spinning NMR spectroscopy to reveal the latent phenotype associated with superoxide dismutase (*sod*-1) and catalase (*ctl*-1) *C. elegans* mutations (both involved in the elimination of radical oxidative species). These two silent mutations are significantly discriminated from the wild-type strain and from each other, providing additional insight into genetic changes.

The genome of the nematode is fully sequenced and the expressed worm proteome shares greater than 83% identifiable homology with human genes (Lai, Chou, Ch'ang, Liu, and Lin, 2000). It also affords a model of primary mitochondrial dysfunction that provides insight into cellular adaptations that accompany mutations in nuclear genes encoding mitochondrial proteins. Falk et al. (2008) applied metabolomics to mutants of the mitochondrial respiratory chain where they showed that primary mitochondrial disease is associated with gene expression alterations in multiple metabolic pathways. Specific pathways that were significantly upregulated in primary mitochondrial respiratory chain disease included those involved in oxidative phosphorylation and the tricarboxylic acid cycle, and those participating in carbohydrate, amino acid, and fatty acid metabolism.

Von Reuss et al. (2012) performed chemical biology-based nematode metabolomics where they used LC/MS-based targeted metabolomics to investigate the biosynthesis of the ascarosides (a family of small molecule signals based on the dideoxy sugar ascarylose and additional building blocks from lipid and amino acid metabolism) in *C. elegans*. Furthermore, Izrayelit et al. (2012) demonstrated that minute changes in ascaroside structures can dramatically affect their signaling properties, and ascaroside biosynthesis and functions are sex-specific.

The *C. elegans* metabolomics technique has also been applied to toxic assessment and toxicological screening. Hughes et al. (2009) performed NMR spectroscopy and UPLC/MS-based metabolomic analyses to investigate the metabolomics responses of nematode to casmium. Their results imply that the main physiological responses to cadmium are independent of the metallothionein status (at least in phytochelatin synthase-normal nematodes) and result in an increased production of phytochelatins by altered flux through the methionine trans-sulfonation pathway. Jones and coworkers (2012) showed that ^{1}H NMR spectroscopy and GC/MS-based metabolomics can be used in conjunction with multivariate statistics to examine the metabolic changes in the nematode following exposure to different concentrations of the heavy metal nickel (the pesticide chlorpyrifos) and their mixture. This work demonstrated the versatility of a combination of *C. elegans* and metabolomics as a functional genomic tool that could form the basis for a rapid and economically viable toxicity test for the molecular effects of pollution/toxicant exposure.

3.4.2.1 *Materials*

The wild type (Bristol N2) and mutants of *C. elegans* strains used in the sample study were obtained from the Caenorhabditis Genetics Center (University of Minnesota, Minneapolis, MN, USA).

Nematodes are grown using standard conditions at 20°C on plates seeded with *Escherichia coli* strain OP50 with the conditions outlined in Strange, Christensen, and Morrison (2007).

To ensure that all the worms are at the same developmental stage, a synchronized population should be produced using the egg preparation technique outlined by Strange et al. (2007) in which gravid adults are bleach-prepped to obtain L1 larvae. Aliquots of approximately 2,000 L1 larvae are transferred to appropriately dosed nematode growth medium (NGM) agar plates. L1 larvae are then allowed to grow until adult stage at 20°C before synchronizing again by bleaching. The resultant larvae (derived from lifetime exposed parents) are then transferred to freshly dosed 90-mm diameter NGM plates and allowed to grow to the L4 stage.

Sample preparation of nematode for ^{1}H high-resolution magic angle spinning NMR (^{1}H HRMAS-NMR) analysis (Blaise et al., 2007):

1. Fix the worms with 3.7% formaldehyde in M9 saline buffer (3 g of KH_2PO_4, 6 g of Na_2HPO_4, 5 g of NaCl, 1 mL of 1 M $MgSO_4$, H_2O to 1 L) for 30 min at room temperature, then wash three times with water.
2. Perform another wash using D_2O to provide a field-frequency lock signal for NMR experiments.

3. Fill a 4-mm HRMAS rotor with Kel-f inserts with a population of 1,000 worms, restricting the effective sample volume to a 12-µL sphere.
4. Use a speed vacuum engine to remove D_2O surplus, and perform NMR acquisition on the same day.

Sample preparation of nematode for ^1H NMR and GC/MS analyses (Jones et al., 2012):

1. Pellet the worms collected from the large plates (approximately 2,000 per plate) by centrifugation at $400 \times g$. Add 600 µL of methanol:chloroform (2:1, v/v) mixture and sonicate for 15 min.
2. Add two aliquots (200 µL) of both chloroform and water, and centrifuge the samples further for 20 min.
3. This results in the formation of aqueous and organic layers, in which the layers are transferred to separate microcentrifuge tubes. Dry the aqueous layer overnight in a Concentrator 5301 evacuated centrifuge (Eppendorf, Histon, UK) and analyze via NMR and GC/MS.
4. Dry the lipid fraction overnight in air and analyze using GC/MS. (This is done because NMR does not detect lipids well.)
5. For 1H NMR analysis, rehydrate the dried extracts from the extraction step in 500 µL of D_2O in order to provide a deuterium lock for the NMR spectrometer.
6. Because NMR chemical shifts are sensitive to pH, add 100 µL of D_2O buffered in 240 mM sodium phosphate, pH 7.0, containing 0.25 mM sodium (3-trimethylsilyl)-2,2,3,3-tetradeuteriopropionate (TSP).
7. For aqueous fraction GC/MS analysis, derivatize the aqueous metabolites via methoximation and trimethyl-silylation prior to analysis.
8. Evaporate 150 µL of sample from the aqueous extract previously analyzed by ^1H NMR spectroscopy to dryness in an evacuated centrifuge.
9. Add 30-µL aliquot of methoxyamine hydrochloride (20 mg mL^{-1}) in pyridine to the dried extract, then mix with a vortex mixer for 30 s before incubating for 17 h at room temperature.

10. Add 30 µL of N-methyl-N-(trimethylsilyl)triluoroacet-amide (MSTFA), mix with vortex for 30 s, and incubate for 1 h.
11. Volume the samples up to 500 µL with hexane prior to analysis via GC/MS.

Sample preparation of nematode for ¹H NMR, GC/MS, and LC/MS analyses (Geier, Want, Leroi, and Bundy, 2011):

1. Split the frozen nematode pellet into two 2-mL tubes, each prefilled with ~100-µL 0.1-mm SiO_2 beads and precooled on dry ice.
2. Add 1 mL of extraction solvent (ice-cold 80% v/v methanol solution in water or an ice-cold monophasic methanol:chloroform:water (2:1:1, v/v/v) modified Bligh and Dyer extraction (1959)) to the frozen pellet.
3. Before the pellet thaws, perform the "destruction and extraction" protocol by bead beating using a FastPrep 12 (MP Biomedicals, Cambridge, UK) operated at 6.5 m s^{-1} for 2 × 30 s with a 1-min ice-cooling step.
4. After extraction, centrifuge all extracts at 16,000 × g for 5 min, then volume up to 2 mL with the above-mentioned extraction solvent.
5. After aliquoting for GC/MS, LC/MS, and NMR in a ratio of 1:1:8, dry the supernatants overnight in a vacuum sample concentrator at room temperature.
6. Keep all samples at −80°C until analysis.

Note: Both 80% methanol and chloroform/methanol combinations give acceptable results (with very similar distributions of reproducibility) across all three analytical platforms. However, the use of 80% methanol rather than chloroform/methanol for extraction is recommended because it is simpler (and, therefore, likely to be more robust for high-throughput studies in particular).

3.4.3 Working with Danio rerio

The zebra fish (*Danio rerio*) has become one of the most widely studied eukaryotes for biological, behavioral, and biomedical research, particularly for studies on embryogenesis, organogenesis, and general vertebrate development (Meyer, Biermann, and Orti, 1993). Because of considerable progress in zebra fish genetics and genomics over the past few years, the zebra fish system has provided many useful tools for studying basic biological processes. In addition, they are easily bred in large numbers with a relatively low maintenance cost compared to rodents. Also, the small size of this fish makes it highly suitable for whole organism assessment of changes in global gene expression caused by chemicals and drugs. Therefore, the zebra fish is now gaining popularity as a model organism for research into disease and drug discovery (Stern and Zon, 2003).

Hayashi et al. (2009) successfully demonstrated the use of metabolomics as a novel strategy to understand the dynamic changes involved in early vertebrate development particularly the embryogenesis in zebra fish. The metabolome during the different stages of embryogenesis exhibited dynamic changes, indicating that the types and quantities of metabolites are correlated with biological activities during development inasmuch as they could successfully predict embryonic stages using metabolomic information. On the basis of the robust correlation between the metabolome and embryogenesis, they concluded that the metabolome can be used as a fingerprint for a particular developmental process.

Soanes and coworkers (2011) highlighted the merger of transcriptomics and metabolomics technologies to study zebra fish embryogenesis. In this study, they combined information from three bioanalytical platforms, namely DNA microarrays, ^{1}H NMR, and direct-infusion mass spectrometry-based metabolomics, to identify and provide insights into the organism's developmental regulators. Their data analysis revealed that changes in transcript levels at specific developmental stages correlate with previously published data with over 90% accuracy. In addition, transcript profiles exhibited trends that were similar to the accumulation of metabolites over time.

Ong, Chor, Zou, and Ong (2009) performed a multiple platform approach, incorporating ^{1}H NMR, GC/MS, and LC/MS to study the biochemical profiles of livers from female and male zebra fish. Their findings showed that ^{1}H NMR provided comprehensive information on glucose, amino acids, pyruvate, and other biochemical constituents of the zebra fish liver. GC/MS spectrometry was able to analyze cholesterol, as well as saturated and unsaturated fatty acids. LC/MS spectrometry was ideal for the analysis of lipids and phospholipids. This multiple techniques approach showed significant differences in the liver of female and male zebra fish. The overall findings suggested that this multiplatform approach offers comprehensive coverage of a metabolome as well as providing valuable insight toward understanding the different biochemical profiles of a biosystem.

3.4.3.1 Materials

The Zebrafish International Resource Center (ZIRC) supplies wild-type, mutant, and transgenic zebra fish to the international research community (Varga, 2011).

Male and female fish (15–20 fish per tank) are kept in separate tanks at a density of 1 fish per 200 mL at 28.5°C (±0.5°C) and under a regulated pH (between 7.3 and 7.4) on a 14:10 h light/dark (L/D) cycle.

The fish are fed twice a day with commercial feed (Zeigler Bros., Inc., USA).

For embryo collection, in the evening, two mating pairs are placed in each mating chamber and divided based on sex. Dividers are pulled within an hour of the lights turning on the following morning to allow for developmental stage matched embryos.

Fertilized eggs are collected immediately after spawning, and the embryos are washed several times in system water, maintained under a more consistent temperature and on a L/D cycle matching the adults, and staged by hpf according to the standard morphological criteria.

Embryos are collected at the various time points, immediately frozen using liquid nitrogen for quenching and stored at −80°C.

For adult fish collection, the fish are removed and kept in a separate tank for a week, water is changed on alternate days, and the fish are starved for 24 h prior to sampling. Upon excision, livers are immediately snap-frozen in liquid nitrogen and subsequently lyophilized overnight.

Sample preparation of zebrafish embryo for GC/MS analysis (Hayashi et al., 2009):

1. Collect 50 zebra fish embryos at various time points in the earliest stage of development.
2. Mix the embryos with 1,000 µL of 80% methanol (diluted in water) with 30 µL of ribitol solution (20 mg mL^{-1}) added as an internal standard.
3. Homogenize the embryos with a MM 301 mixer mill (Retsch GmbH, Haan, Germany), and incubate the homogenate for 30 min at 37°C and before centrifugation at 16,000 × g for 3 min at 4°C.
4. Transfer an 800-µL aliquot of the supernatant to an Eppendorf tube with a pierced cap, and dry the sample in a vacuum centrifuge dryer.
5. For oximation, add 50 µL methoxyamine hydrochloride in pyridine (20 mg mL^{-1}) to the dried hydrophilic crude extract, and incubate in a Thermomixer comfort (Eppendorf) at 1,200 rpm for 90 min at 30°C.
6. Add 50 µL of N-methyl-N-(trimethylsilyl)triluoroacetamide (MSTFA) and incubate at 37°C for 30 min in order to induce the silylation reaction before injecting to the GC/MS machine.

Sample preparation of zebra fish liver for GC/MS, LC/MS, and ^{1}H NMR analyses (Ong et al., 2009):

1. Freeze-dry the liver collected from individual fish and extract with 0.5 mL of methanol:chloroform (1:3, v/v).
2. After centrifugation at 15,000 × g for 5 min, lyophilize the supernatant.
3. Keep the dried sample (lipid fraction) at −30°C prior to analysis using GC/MS and LC/MS.
4. For ^{1}H NMR analysis, extract the aqueous fraction of the liver tissue with 0.5 mL acetonitrile:water (1:1, v/v).
5. After centrifugation, lyophilize the supernatant.
6. Subsequently, extract the residues with 0.5 mL of methanol:chloroform (1:3, v/v), and obtain the lipid fraction after drying the supernatant.
7. Keep all samples at −30°C until analysis.

3.5 Clinical Samples

With the recent breakthrough in metabolomics technologies, application of metabolomics has been increasing in the medical field. Metabolic profiling is defined as the identification and quantitation of the compounds in the metabolome and is applied to define metabolic changes related to genetic differences, environmental influences, and disease or drug perturbations. Clinical metabolomics has two major purposes in medicine: to acquire knowledge on the mechanisms of drug action or the disease itself, and for biomarker detection and disease diagnosis. In particular, studies on biomarker discovery using metabolomics techniques have attracted attention worldwide as a practical application for disease diagnostics. Moreover, metabonomics can also be useful for determining the therapeutic potential of metabolites whose levels are altered in a particular disease state (Arakaki, Skolnick, and McDonald, 2008).

The next sections introduce representative sample preparation protocols for urine, serum, or organ and tissue for use in the medical field.

3.5.1 Urine

Urine is a very popular biofluid for metabolomic investigation for the following reasons: (1) its collection is noninvasive, (2) the complex metabolic nature of the fluid, and (3) the ability to collect multiple samples over a period of time. There are numerous reports on the metabolomics-based studies using urine as the sample for medical applications. For example, metabolomics has been used to study hepatocellular carcinoma (HCC), which is the sixth most prevalent malignant tumor in the world and is ranked as the third most lethal form of cancer, causing more than half a million deaths annually in the world (Bosch, Ribes, Díaz, and Cléries, 2004). Wu et al. (2009) studied metabolite profiling of human urine samples from HCC patients and healthy subjects using GC/MS and established a PCA-based (principal components analysis) diagnostic model from 18 metabolic biomarkers that can distinguish HCC from normal subjects. Multivariate statistics with PCA and receiver–operator characteristic curves (ROC) of the diagnostic model yielded a separation between the two groups with an area under the curve value of 0.9275. Therefore, they concluded that this noninvasive technique of identifying HCC biomarkers from urine may have clinical utility.

Bladder cancer, on the other hand, is the fourth most common cancer in American men and accounts for more deaths annually in women than cervical cancer (Jemal, Siegel, Xu, and Ward, 2010). Although alterations in xenobiotic metabolism are considered causal in the development of bladder cancer, the precise mechanisms involved are poorly understood. Putluri and coworkers (2011) performed an LC/MS-based metabolomics study for discovery of human urinary biomarker for early detection and staging of bladder cancer. Their findings identified candidate diagnostic and prognostic markers for bladder cancer and highlighted mechanisms associated with the silencing of xenobiotic metabolism.

There are a number of methods in use for urine collection in animal experiments. Forced urination by compression of the lumbar region is a typical method for sampling small volume urine, and it has been usually chosen for sampling fresh urine, which is clinically tested just after urination. On the other hand, when a large volume of urine is needed, pool urine is obtained: urine is pooled by housing animals in metabolism cages for several hours in animal experiments. There is a question that sampling methods or periods can be determined only based on required sample volume for analysis on designing animal experiments. In addition, urine is usually pooled in metabolism cages at animal room temperature during sampling (i.e., several hours). Another question is that excrete metabolites are stable at room temperature during pooling urine. Bando et al. (2010) examined the influences of urine sample collection and handling procedures on the metabolic profiles, and attempted to propose appropriate sampling methods such that pathophysiological responses can be found in distinction from biological/technical variation.

3.5.1.1 *Materials*

3.5.1.1.1 *Human Urine Collection*

The human urine samples used should be in accordance with the guidelines used in assessing hospital autonomy in each country, and written informed consent should be obtained from all subjects.

In general, the patients are diagnosed by microscopy, biopsy, or surgical resection and classified using International Classification of Diseases (ICD; Organization, 2013).

Each urine sample was obtained from patients and healthy volunteers after fasting in the early morning.

Following collection, urine specimens are centrifuged at 1 h at $800 \times g$ for 10 min at room temperature and the supernatant is stored at $-80°C$ until sample extraction.

3.5.1.1.2 *Animal Urine Collection*

Animal study is conducted in compliance with the "Law for Partial Amendments to the Law concerning the Protection and Control of Animals" in each country.

Two Male Crl:CD(SD) rats per cage are housed in aluminum wire mesh cages or rats are housed individually in metabolic cages during the collection of urine, and maintained in an air-conditioned animal room (temperature: $24.0 \pm 2.0°C$, relative humidity: $55.0 \pm 15.0\%$) with a 12-h light–dark cycle and more than 10 air changes per hour.

The rats receive pelleted basal diet and tap water ad libitum, however, diet is removed during urine collection.

Pooled urine samples under iced conditions are thawed in a refrigerator after the collection period, all urine samples are centrifuged at $540 \times g$ for 5 min at $5°C$, and portion of supernatant is used for measurement of creatinine concentration. Creatinine concentrations of urine samples are measured by sarcosine oxidase/peroxidase method (BM-1650, JOEL, Japan).

The remaining urine samples are divided into subsamples and frozen rapidly using liquid nitrogen and stored at $-80°C$ until sample extraction.

Note: Metabolic profiles of urine with GC/MS vary dramatically depending on urine pooling period and tube conditions, underscoring the importance of determining appropriate sampling periods in consideration of diurnal effects and targets of effect/toxicity, and suggesting it would be preferable to keep tubes in metabolic cages under iced conditions for urine sampling (Bando et al., 2010).

Sample preparation of human urine for LC/MS analysis (Putluri et al., 2011):

1. Measure the osmolarity of all urine samples using the MULTI-OSMETTE 2430 osmometer (Precision Systems Inc., Natick, MA, USA) and calibrate the values using standards as per manufacturer's instruction. The osmolarity of the urine samples for examination should be restricted to 140 to 400 milliosmoles/L which is achieved by measuring out a defined volume of the urine prior to extraction.
2. Introduce an equimolar mixture of 4 internal standard compounds (^{15}N-tryptophan, d_4-thymine, ^{15}N-N-acetyl aspartic acid, and d_5-glutamic acid) into the specimen and dry the mixture under vacuum.
3. Prior to analysis, resuspend all samples in an equal volume of water: acetonitrile (98:2, v/v) with 0.1% formic acid prior to LC/MS analysis.

Sample preparation of animal urine for GC/MS analysis (Bando et al., 2010):

1. Mix 100 µL of urine sample with 500 µL of methanol:chloroform:water (2.5:1:1, v/v/v) and 180 µL of ribitol (200 mg mL^{-1}) (as internal standard).
2. Shake the samples for 30 min at 37°C and centrifuge at $16,000 \times g$ for 3 min at 4°C.
3. Transfer 400 µL of the supernatant to a new Eppendorf tube with a pierced cap. Evaporate the samples under vacuum using a centrifugal concentrator at room temperature for approximately 2 h, followed by freeze-drying overnight before derivatization.

4. For oximation, add 100 µL of methoxyamine hydrochloride in pyridine (20 mg mL^{-1}) to the dried hydrophilic crude extract, and incubate in a Thermomixer comfort (Eppendorf) at 1,200 rpm for 90 min at 30°C.
5. For silylation, add 50 µL of N-methyl-N-(trimethylsilyl) triluoroacetamide (MSTFA) and incubate at 37°C for 30 min before injecting to the GC/MS machine.
6. Finally, normalize the intensities of each peak obtained by GC/MS by creatinine concentrations obtained from sarcosine oxidase/peroxidase measurements.

3.5.2 Serum

Blood (serum or plasma), similar to urine, is one of the most frequently studied samples for a number of reasons: (1) blood sample collection is minimally invasive compared to the collection of cerebrospinal fluid (CSF) and tissues and (2) blood exhibits the functions and phenotypes of many different parts of the body in a single sample. Thus, blood metabolomics is defined as the "metabolic footprint" of tissue metabolism (Kell et al., 2005). The Human Metabolome Database (HMDB) is currently the most complete and comprehensive curated collection of human metabolite and human metabolism data in the world (Wishart et al., 2007). To date, the HMDB curates a vast number of identified and quantified metabolites including 309 in CSF, 1,122 in serum, 458 in urine, and approximately 300 from other tissues and biofluids. The HMDB data indicate that serum includes a great deal of metabolite information compared with other biofluids.

Numerous studies have been performed to identify cancer-specific biomarkers in human patients. For example, Nishiumi et al. (2010) identified 18 serum biomarkers of pancreatic cancer, which is difficult to detect and diagnose early. As a followup, the same group applied metabolomics techniques to establish new screening methods for early diagnosis of colorectal cancer (Nishiumi et al., 2012). The metabolites whose levels displayed significant changes were subjected to multiple logistic regression analysis using the stepwise variable selection method, and a colorectal cancer prediction model was established. The prediction model was composed of multiple biomarkers (i.e.,

2-hydroxybutyrate, aspartic acid, kynurenine, and cystamine). Although the training set included patients with early stage colorectal cancer, such as stage 0 or stage 1, the metabolomics-based prediction model displayed high AUC (0.9097), sensitivity (85.0%), and accuracy (85.0%) values, which were higher than those of conventional serum tumor markers (CA19-9 and CEA). They concluded that GC/MS-based serum metabolomics could be used as a novel method for colorectal cancer screening tests. Additionally, the pathogenesis of colorectal cancer seems to lead to alterations in the levels of a variety of serum metabolites, although these fluctuations range from small to large.

More recently, large-scale studies of various human diseases were carried out worldwide. Dunn et al. (2011) proposed an experimental workflow for long-term and large-scale metabolomic studies involving thousands of human samples with data acquired from multiple analytical batches over many months and years. They also described protocols for serum- and plasma-based metabolic profiling for GC/MS and LC/MS. In particular, procedures for biofluid collection, sample preparation, data acquisition, data preprocessing, and quality assurance were described in more detail. Accordingly, we predict that their protocol will be widely used as a standard method for the biomarker discovery with long-term and large-scale metabolomic studies.

On the other hand, there are few methods available for plasma collection in animal experiments. From plasma, "snapshots" of metabolic profiles and onset of toxicity can be observed, thereby predicting toxicity. Although a few factors (e.g., fasting, sampling site of bleeding, stress, and anticoagulant) are known to affect clinical biochemistry data (Neptun, Smith, and Irons, 1985), influences of collection procedures on metabolic profiles have rarely been investigated. Bando et al. (2010) examined the effect of plasma sample collection and handling procedures on the metabolic profiles of rats. Metabolic profiles of plasma differed depending on blood sampling sites. Anesthesia was not effective in reducing individual variation, although the anesthesia was beneficial in reducing discomfort in rats. In GC/MS-based metabolomics studies, they concluded that EDTA be used as anticoagulant in plasma sample preparation, because peaks derived from heparin might overlap with endogenous metabolites, which may induce intersample variation.

3.5.2.1 Materials

3.5.2.1.1 Human Blood Collection

The human serum or plasma samples are used in accordance with the guidelines used in assessing hospital autonomy in each country, and written informed consent is obtained from all subjects.

In general, the patients are diagnosed by microscopy, biopsy, or surgical resection and classified using International Classification of Diseases (ICD).

Blood samples are obtained from patients and healthy volunteers after fasting in the early morning.

Blood (typically 10 mL) is collected into lithium heparin tubes (e.g., BD vacutainer, Li-heparin) to give plasma, or left on ice for a minimum of 1 h to coagulate (e.g., BD vacutainer, no additive). After centrifugation at 2,500 × g for 15 min at 4°C, this results in serum.

Samples are then immediately divided into aliquots (e.g., 0.5 mL) in cryovials and frozen at −80°C until sample preparation procedures are carried out.

3.5.2.1.2 Animal Plasma Collection

Animal study is conducted in compliance with the "Law for Partial Amendments to the Law concerning the Protection and Control of Animals" in each country.

The rats received pelleted basal diet and tap water ad libitum, however, diet is removed 8 h before plasma collection.

Plasma samples are drawn via the jugular vein or abdominal aorta under isoflurane anesthesia from rats.

Blood samples collected are treated with anticoagulants, then centrifuged at 2,150 × g for 10 min at 5°C, and supernatants (plasma) are divided into aliquots in cryovials and frozen rapidly using liquid nitrogen and stored in a deep freezer (−80°C) until sample extraction.

Note: A number of anticoagulants are available, including potassium EDTA, citrate, and lithium heparin. In NMR analysis, both citrate and EDTA can interfere with subsequent metabolic profiling, either by introducing interfering peaks or, in the case of citrate, by obscuring the endogenous analyte (Beckonert et al., 2007). For this reason, the use of lithium heparin is recommended for preparing plasma samples for general analysis. However, in GC/MS analysis, EDTA should be used as anticoagulant in plasma sample preparation, because peaks derived from heparin might overlap with endogenous metabolites (Bando et al., 2010). Consequently, the effect of anticoagulants for each analytical system should be checked first.

Sample preparation of human serum for GC/MS analysis (Nishiumi et al., 2010):

1. Allow the serum samples to thaw on ice at 4°C for 30–60 min.
2. Mix the serum sample (50 µL) with 250 µL of methanol:chloroform:water (2.5:1:1, v/v/v) containing 6 µL of 2-isopropylmalic acid (1 mg mL^{-1}) as internal standard.
3. Shake the samples for 30 min at 37°C and centrifuge at 16,000 × g for 5 min at 4°C.
4. Transfer the supernatant (250 µL) to a new Eppendorf tube and add 200 µL of distilled water to the tube.
5. After mixing, centrifuge the solution at 16,000 × g for 5 min at 4°C, and transfer 250 µL of the resultant supernatant to a clean tube, before lyophilizing using a freeze-dryer.
6. For oximation, add 40 µL of methoxyamine hydrochloride in pyridine (20 mg mL^{-1}) to the dried hydrophilic crude extract, and incubate in a Thermomixer comfort (Eppendorf) at 1,200 rpm for 90 min at 30°C.
7. For silylation, add 20 µL of N-methyl-N-(trimethylsilyl) triuoroacetamide (MSTFA) and incubate at 37°C for 30 min, before injecting to the GC/MS machine.

Sample preparation of human plasma/serum for GC/MS and LC/MS analyses (Dunn et al., 2011):

1. Allow the plasma/serum samples to thaw on ice at 4°C for 30–60 min.
2. Extract an aliquot of plasma/serum sample (400 µL) with 1,200 µL of methanol and 200 µL of internal standards including d_2-malonic acid, d_4-succinic acid, d_5-glycine, d_4-citric acid, $^{13}C_6$-D-fructose, d_5-L-tryptophan, d_4-L-lysine, d_7-L-alanine, d_{35}-stearic acid, d_5-benzoic acid, d_{15}-octanoic acid (each 0.167 mg mL^{-1}).
3. Mix the samples thoroughly on a vortex mixer for 15 s, and precipitate the protein pellet in a centrifuge operating at room temperature for 15 min at $15,800 \times g$.
4. Transfer 370 µL of the supernatant into four separate new tubes (GC/MS, LC/MS positive ion mode, LC/MS negative ion mode, and reserve) and lyophilize each sample in a centrifugal vacuum evaporator at room temperature (no heating) for 18 h.
5. For GC/MS analysis, after the dried samples are lyophilized using a freeze-dryer for 1 h, the freeze-dried samples are set on a Dri-Block heater and allowed to reach a setpoint temperature of 80°C.
6. For oximation, add 50 µL methoxyamine hydrochloride in pyridine (20 mg mL^{-1}) to the dried extract. Mix the reaction solution thoroughly for 15 s and heat in the Dri-Block at 80°C for 15 min.
7. Add 50 µL of N-methyl-N-(trimethylsilyl)triluoroacetamide (MSTFA) and heat again at 80°C for 15 min.
8. Remove the samples from the block heater and allow to cool for 5 min. Finally, add 20 µL of the working retention index solution (docosane, nonadecane, decane, dodecane, and pentadecane).
9. For LC/MS analysis, add 100 µL of water to the dried plasma/serum extract.
10. After mixing, centrifuge the solution at $15,800 \times g$ for 15 min, and transfer 90 µL of the resultant supernatant to a clean analytical vial.

3.5.3 Organ and Tissue

Metabolomic analysis of tissues from patients or animal models is a powerful tool for studying the abnormal metabolisms of diseases, and it can provide information about the metabolic modifications and the upstream regulative mechanisms in diseases (Griffin and Nicholls, 2006). More important, the systemic metabolic characteristics of tissues could provide opportunities for exploring novel diagnostic markers or therapeutic targets for clinical applications.

Sreekumar et al. (2009) studied the GC/MS- and LC/MS-based metabolomics of the metabolic mechanism in prostate cancer progression. Tissue samples were derived from benign adjacent prostate ($n = 16$), clinically localized prostate cancer ($n = 12$), and metastatic prostate cancer ($n = 14$) patients. In total, metabolomic profiling of the tissues quantitatively detected 626 metabolites (175 named, 19 isobars, and 432 metabolites without identification), of which 82.3% (515 out of 626) were shared by the three diagnostic classes. Notably, there were 60 metabolites found in clinically localized prostate cancer or metastatic tumors but not in benign prostate. These unbiased metabolomic profiles were able to distinguish benign prostate, clinically localized prostate cancer, and metastatic disease. Sarcosine, an N-methyl derivative of the amino acid glycine, was identified as a differential metabolite that was highly increased during prostate cancer progression to metastasis. By profiling the metabolomic alterations of prostate cancer progression, Sreekumar and coworkers revealed sarcosine as a potentially important metabolic intermediary of cancer cell invasion and aggressiveness.

Although alterations in xenobiotic metabolism are considered causal in the development of bladder cancer, the precise mechanisms involved are poorly understood. Putluri et al. (2011) studied the mechanisms associated with the silencing of xenobiotic metabolism in bladder cancer development and progression. They used LC/MS to measure over 2,000 compounds in 58 clinical specimens, identifying 35 metabolites that exhibited significant changes in bladder cancer. This metabolic signature distinguished both normal and benign bladder from bladder cancer. Subsequent enrichment-based bioprocess mapping revealed alterations in phase I/II metabolism and suggested a possible role in DNA methylation for perturbing xenobiotic metabolism in bladder cancer. Putluri and coworkers highlighted the utility of evaluating metabolomic profiles of bladder cancer to gain insights into bioprocesses perturbed during bladder cancer development and progression.

The Mouse Multiple Tissue Metabolome Database (MMMDB) is a freely available metabolomic database containing a collection of metabolites measured from multiple tissues from single mice (Sugimoto et al., 2012). Currently, data from cerebra, cerebella, thymus, spleen, lung, liver, kidney, heart, pancreas, testis, and plasma are provided. MMMDB contains the concentrations of a large number of metabolites simultaneously profiled using capillary electrophoresis time-of-flight mass spectrometry (CE-TOFMS) in a nontargeted manner. MMMDB is very useful for capturing the holistic overview of large metabolomic pathway.

3.5.3.1 Human Tissue Collection

The human tissue samples are used in accordance with the guidelines used in assessing hospital autonomy in each country, and written informed consent is obtained from all subjects.

In general, the patients are diagnosed by microscopy, biopsy, or surgical resection and classified using International Classification of Diseases (ICD).

Tissue samples obtained from patients are immediately frozen in liquid nitrogen and stored at −80°C until sample extraction.

3.5.3.2 Animal Tissue Collection

Animal study is conducted in compliance with the "Law for Partial Amendments to the Law concerning the Protection and Control of Animals" in each country.

The rats received pelleted basal diet and tap water ad libitum, however, diet is removed 8 h before plasma collection.

Rat/mouse tissue samples collected are frozen rapidly using liquid nitrogen and stored in a deep freezer (−80°C) until sample extraction.

Sample preparation of human tissue for LC/MS analysis (Putluri et al., 2011):

1. Homogenize and extract 25 mg of frozen tissue in ice cold methanol:water (4:1, v/v) mixture containing equimolar mixture of 11 standard compounds (epibrassinolide, d_3-testosterone, $^{15}N_1$-anthranilic acid, zeatine, jasmonic acid, gibberelic acid, d_4-estrone, $^{15}N_1$-tryptophan, d_4-thymine, $^{13}C_1$-creatinine, and $^{15}N_1$-Arginine).

2. Subject the samples to separation of the polar and nonpolar phases by sequential addition of ice cold chloroform and water in 3:1 ratio (v/v) and separation of the organic (methanol and chloroform) and aqueous solvents.

3. Deproteinize the aqueous extract using a 3-KDa molecular filter (Amicon Ultracel-3K Membrane, Millipore Corporation, Billerica, MA) and dry the filtrate containing metabolites under vacuum.

4. Resuspend the dried extract in an equal volume of injection solvent composed of methanol:water (1:1, v/v) with 0.2% acetic acid before subjecting to LC/MS.

Sample preparation of animal tissue for hydrophilic metabolites analysis with CE/MS (Sugimoto et al., 2012):

1. Immediately plunge 50 mg of frozen tissue in 500-µL methanol containing internal standards (20 µM of methionine sulfone and D-camphor-10-sulfonic acid) and homogenize for 3 min to inactivate the enzymes.

2. Add 200 µL of Milli-Q water and 500 µL of chloroform, and mix thoroughly.

3. Centrifuge the solution at 4,600 $\times$ g for 15 min at 4°C and centrifugally filter 450 µL of the upper aqueous layer through a Millipore 5-kDa cutoff filter (Human Metabolome Technologies Inc., Tsuruoka, Japan) to remove large molecules.

4. Lyophilize the filtrate and dissolve in 50 µL Milli-Q water containing the reference compounds (200 µM each of 3-aminopyrrolidine and trimesate) prior to CE-TOFMS analysis.

3.6 Food Samples

Food is composed of thousands of components resulting from the complex metabolism of plants and animals. Furthermore, the physical and chemical interactions of all compounds present in food directly affect the final attributes of the product such as sensory characteristics, quality, shelf life, and safety. Therefore, in order to understand food processes and systems, it is imperative to identify the food components and interactions that are responsible for the characteristics of food samples (Cevallos-Cevallos and Reyes-De-Corcuera, 2011). Metabolomics, the interdisciplinary study that involves the quantitative profiling of metabolites in a system, provides a standard and exhaustive approach for identifying and quantifying food components that influence the characteristics of the food product (Putri et al., 2013).

Applications of metabolomics technology in food science include: (1) informative analyses to characterize and identify compounds of interest, (2) prediction of quantitative functional value of food by means of multivariate analysis using metabolome data as explanatory variable (herein termed "predictive metabolomics"), and (3) comparative metabolomics to determine the metabolites responsible for classification of samples by type or for discriminatory purposes (Putri, Bamba, and Fukusaki, 2013). Informative metabolomic analyses are often used to identify and quantify compounds of interest to obtain basic information useful in the development of metabolite databases (such as the *E. coli*, yeast, and human metabolome databases) and the discovery of pathways, novel bioactive compounds, and biomarkers (Wishart et al., 2007; Cevallos-Cevallos, Reyes-De-Corcuera, Etxeberria, Danyluk, and Rodrick, 2009; Guo et al., 2013; Jewison et al., 2012). Predictive metabolomics is used to calculate a dependent variable that is difficult to measure based on statistical models created from a metabolite profile as exemplified by the metabolite-based models for prediction of green tea sensory quality (Ikeda, Kanaya, Yonetani, Kobayashi, and Fukusaki, 2007; Pongsuwan et al., 2007). On the other hand, discriminative metabolomics can establish the differences in metabolites between sample populations that can be used to develop new food plant cultivars or classify food (e.g., wine) based on variety and production area (Cevallos-Cevallos and Reyes-De-Corcuera, 2011; Khan et al., 2012; Son et al., 2008).

This section introduces representative sample preparation protocols for fermented products such as cheese, soy sauce, and sake as well as crop plants such as watermelon, green tea, and coffee.

3.6.1 Working with Fermented Products

3.6.1.1 Cheese, Soy Sauce, and Sake

Fermented foods, produced or preserved by the action of microorganisms, play important roles in the diet of many nationalities as they are linked to their cultural identity, local economy, and gastronomical delight (Hui and Evranuz, 2012). Fermentation typically refers to the conversion of sugar or other carbohydrates to alcohols and carbon dioxide or organic acids using yeast, bacteria, or both under anaerobic conditions. Some examples of fermented products resulting from fermentation include cheese, beer, sake, soy sauce, and sauerkraut.

Recently, metabolomics has emerged as an essential tool for the evaluation of quality and safety of food samples because multiple factors can be analyzed simultaneously (Putri et al., 2013). In fact, metabolomics has been utilized for the analysis of fermented products such as wine (Vaclavik, Lacina, Hajslova, and Zweigenbaum, 2011), beer (Intelmann et al., 2011; Rodrigues, Barros, Carvalho, Brandão, and Gil, 2011), cheese (Ochi, Naito, Iwatsuki, Bamba, and Fukusaki, 2012), soy sauce (Yamamoto et al., 2012), sake (Sugimoto et al., 2009), and meju (fermented soybean; Ko, Kim, Hong, and Lee, 2010). To evaluate the sensory attributes of food samples, it is important to analyze low-molecular–weight hydrophilic components that are highly correlated to taste as well as amino acids, organic acids, and sugars. In this case, GC/MS can be used to analyze these components comprehensively in biological samples (Yamamoto et al., 2012).

3.6.1.2 *Materials*

3.6.1.2.1 Cheese Sample Collection

Natural cheese samples (Cheddar, Gouda, Parmigiano, and Reggiano) are collected from various locations including Australia, United Kingdom, The Netherlands, Italy, and Japan.

Samples are stored at −80°C until sample preparation. Cheese samples are frozen in liquid nitrogen and ground until powder form. Freeze-dry the samples prior to sample preparation.

Sample preparation of natural cheeses for GC/MS analysis (Ochi et al., 2012):

1. Place 100 mg of the freeze-dried sample in a 2-mL Eppendorf tube and add 1,000 µL of methanol: chloroform: water (2.5:1:1, v/v/v) and then 60 µL of 0.2 mg mL^{-1} ribitol (as internal standard).

2. Place 5-mm diameter zirconia beads in the tube and suspend in a ball-mill (20 Hz) for 1 min at room temperature followed by sonication for 90 s (3×).

3. Centrifuge the sample at 16,000 × g, for 3 min at 4°C and transfer 800 µL of the supernatant to a new 1.5-mL microfuge tube.

4. Add 400 µL of Milli-Q water and mix using a vortex.

5. Centrifuge the sample at 16,000 × g, for 3 min at 4°C and transfer 500 µL of the supernatant to a new 1.5-mL microfuge tube capped with a pierced cap.

6. Remove the methanol from the extract by using a centrifugal vacuum concentrator at room temperature for approximately 2 h.

7. After evaporation, freeze-dry the extract in a glass bottle at room temperature overnight.

8. For methoxyation reaction, add 100 µL of methoxyamine hydrochloride in pyridine (20 mg mL^{-1}) to the freeze-dried sample and incubate in a Thermomixer at 30°C for 90 min.

9. For silylation reaction, add 50 µL of N-Methyl-N-(trimethylsilyl)trifluoroacetamide to the mixture and incubate in a Thermomixer at 37°C for 90 min.

3.6.1.3 *Materials*

3.6.1.3.1 *Soy Sauce Sample Collection*

Soy sauce samples were purchased from the markets around the world. A total of 24 samples of Japanese, Chinese, and Western soy sauces were used in this study.

Sample preparation of soy sauce for GC/MS (Yamamoto et al., 2012):

1. Add 1 mL of methanol:chloroform:water (2.5:1:1, v/v/v) to 20-µL soy sauce sample in a 1.5-mL tube. Add 60 µL of 0.2 mg mL^{-1} ribitol to the mixture as internal standard.
2. Sonicate the mixture for 1 min at room temperature.
3. Centrifuge the sample at 13,400 × g for 3 min. at 4°C and transfer 900 µL of the supernatant to a new 1.5-mL tube.
4. Add 400 µL of distilled water and vortex before centrifugation at 13,400 × g for 3 min at 4°C.
5. Transfer 900 µL of the polar phase to a new 1.5-mL microtube capped with a pierced cap.
6. Dry the extract under vacuum in a centrifugal concentrator for 2 h and in a freeze-dryer until completely dry.
7. To induce methoxymation reaction, add 50 µL of methoxyamine hydrochloride in pyridine (20 mg mL^{-1}) to the dried hydrophilic extract and incubate in a Thermomixer at 30°C for 90 min.
8. For silylation reaction, add 50 µL of N-Methyl-N-(trimethylsilyl)trifluoroacetamide to the mixture and incubate in a Thermomixer at 37°C for 30 min.
9. Centrifuge the reaction mixture at 13,400 × g for 3 min and transfer the supernatant to a vial for GC/MS analysis.

Note: Soy sauce contains large quantities of lipophilic components from soy, wheat, and sodium chloride that may interfere with GC/MS analysis. Therefore, separation of hydrophilic and lipophilic components should be performed by addition of water following extraction with a mixture solvent (methanol:chloroform:water). Sodium chloride is removed from pyridine solution by centrifugation after derivatization due to its low solubility in pyridine.

3.6.1.4 Materials

3.6.1.4.1 Sake Sample Collection

Forty various types of sake samples, such as ordinary sake and specially designated sake, were purchased from the market (June 2012, Hiroshima, Japan).

Sample preparation of sake (nonvolatile component) for GC/TOFMS (unpublished data):

1. Add 60 µL of 0.2 mg mL^{-1} ribitol (internal standard) to 20-µL sake sample in a 1.5-mL microfuge tube.
2. Dry the mixture under vacuum in a centrifugal concentrator for 2 h.
3. To induce methoxymation reaction, add 100 µL of methoxyamine hydrochloride in pyridine (20 mg mL^{-1}) to the dried hydrophilic extract and incubate in a Thermomixer at 30°C for 90 min.
4. For silylation reaction, add 50 µL of N-Methyl-N-(trimethylsilyl)trifluoroacetamide to the mixture and incubate in a Thermomixer at 37°C for 30 min.
5. Centrifuge the reaction mixture at 13,400 × g for 3 min and transfer the supernatant to a vial for GC/MS analysis.

Sample preparation of sake (volatile component using SBSE method) for GC ("unpublished data"):

1. Dispense 10 mL of sake sample (adjusted to 10% (v/v) ethanol concentration with pure water) into a 10-mL glass headspace vial.
2. Add 2 g of NaCl and 3-octanol (to a final concentration of 0.5 mg L^{-1}) as internal standard.
3. Stir the mixture at 800 rpm for 1 h at room temperature (24°C) using a Twister™ stir bar.
4. Remove the stir bar from the sample and transfer to a glass thermal desorption tube.
5. Place the desorption tube into the Gerstel TDS 2 Thermodesorption unit and thermally desorb the stir bar by heating the TDS 2 unit from 20°C (for 1 min) to 230°C (for 4 min) at 60°C min^{-1}. The trapped components are used for injection into the GC column.

3.6.2 Working with Crop Plants

3.6.2.1 Watermelon, Green Tea, Coffee

A crop plant is a cultivated plant grown on a large scale whose product such as grain, fruit, or vegetable is harvested by humans at some point of its growth stage for food, clothing, or other human uses. Major crops include wheat, rice, corn, cassava, soybeans, and potatoes among others. Traditionally, the quality of food products is assessed by highly trained specialists via sensory evaluation. In the past few decades, considerable effort has been given to the development of a methodology capable of utilizing instrumental data to supplement or replace organoleptic sensory analysis for performing quality analysis of foods and beverages (Putri et al., 2013). Metabolomics has been regarded as a powerful technology to capture the sample complexity and extract the most meaningful elements for quality evaluation of food. Metabolomic analysis has been widely applied for assessing food quality, food safety, and determination of origin, authentication, and varietal differences of various important food crops such as watermelon (Tarachiwin et al., 2007), green tea (Pongsuwan et al., 2007, 2008), and coffee (Jumhawan, Putri, Marwani, Bamba, and Fukusaki, 2013).

3.6.2.2 Materials

3.6.2.2.1 Watermelon Sample Collection

Seven graded watermelons grafted on gourd (Kachidoki2gou) rootstocks and three graded watermelons grafted on pumpkin (Shintosa, V-999, Kagayaki) rootstocks obtained from Hagiwara Farm Co. Ltd., Japan, were used in this study.

Sample preparation of watermelon for NMR analysis (Tarachiwin et al., 2008):

1. Extract the juice from the center and placenta of the watermelon separately and mix using a blender. Filter the juice using a 0.45-μm PTFE membrane (Advantec).

2. Mix 200 µL of supernatant juice with 200 µL of D_2O-containing 3 mM DSS and 200 µL of 0.2 M phosphate buffer to make a 600-µL solution for NMR measurement.

Note: It is recommended to prepare all the samples in 1 day and keep in 4°C prior to analysis.

3.6.2.3 Materials

3.6.2.3.1 Green Tea Sample Collection

Dried tea leaves were collected from various places in Japan and kept at −30°C prior to sample preparation.

Sample preparation of green tea for GC/MS analysis (Pongsuwan et al., 2007):

1. Place 30 mg of dried tea leaves in a 2-mL microfuge tube and freeze-dry using liquid nitrogen.
2. Ground the freeze-dried samples with a Retsch ball mill for 1 m at 20 Hz.
3. Extract the metabolites using methanol:chloroform: water (2.5:1:1, v/v/v) and then 60 µL of 0.2 mg mL^{-1} ribitol (as internal standard).
4. Centrifuge the sample at 16,000 × g for 3 m at 4°C and transfer 900 µL of the supernatant to a new 1.5-mL Eppendorf tube.
5. Add 400 µL of Millipore Milli-Q system-purified water and mix using a vortex.
6. Centrifuge the sample at 16,000 × g for 3 m at 4°C and transfer 400 µL of the polar phase to a new 1.5-mL microfuge tube capped with a pierced cap.
7. Dry the extract in a vacuum centrifuge concentrator at room temperature for approximately 2 h.
8. After evaporation, freeze-dry the extract in a glass bottle at room temperature overnight.
9. For methoxyation reaction, add 50 µL of methoxyamine hydrochloride in pyridine (20 mg mL^{-1}) to the freeze-dried sample and incubate in a Thermomixer at 30°C for 90 m.

10. For silylation reaction, add 100 µL of N-Methyl-N-(trimethylsilyl)trifluoroacetamide to the mixture and incubate in a Thermomixer at 37°C for 90 m.

3.6.2.4 Materials

3.6.2.4.1 Coffee Sample Collection

Coffee samples were obtained from 21 sampling points in three cultivation areas in Indonesia (Java, Sumatra, and Bali).

Samples were roasted in a Probat-Werke von Gimborn Maschinenfabrik GmbH model BRZ 2 (Probat, Rhein, Germany) at 205°C for 10 m to obtain a medium degree of roasting and then were air-cooled for 5 m.

Coffee beans were ground and stored in sealed 50 mL BD Falcon tubes at −30°C with light-shielding prior to analysis.

Sample preparation of coffee for GC/MS analysis (Jumhawan et al., 2013):

1. Put coffee beans in a precooled grinding mill container and grind using a Retsch ball mill for 3 m at 20 Hz.
2. Transfer 15 mg of the coffee bean powder into a 2-mL microfuge tube.
3. Add 1 mL of methanol:chloroform:water (2.5:1:1, v/v/v) and then 60 µL of 0.2 mg mL^{-1} ribitol (as internal standard).
4. Vortex the mixture for 1 m and centrifuge at 16,000 × g for 3 min at 4°C.
5. Transfer 900 µL of the supernatant into a 1.5-mL microfuge tube and add 400-µL Milli-Q water.
6. Vortex the mixture for 1 m and centrifuge at 16,000 × g for 3 m at 4°C.
7. Transfer 400 µL of the aqueous phase to a new 1.5-mL microfuge tube with a screw cap.
8. Remove the solvent by vacuum centrifugation for 2 h, followed by freeze-drying overnight.
9. For methoxyation reaction, add 100 µL of methoxyamine hydrochloride in pyridine (20 mg mL^{-1}) to the freeze-dried sample and incubate in a Thermomixer at 30°C for 90 m.

10. For silylation reaction, add 50 μL of N-Methyl-N-(trimethylsilyl)trifluoroacetamide to the mixture and incubate in a Thermomixer at 37°C for 90 m.

3.7 Summary

In this chapter, we introduce the sample preparation protocols for metabolomics analysis in wide-ranging studies involving microbiology, plant, animal, medical sciences, and food subjects. Inasmuch as the quality and reliability of metabolomics data will invariably depend on the sampling and sample treatment techniques employed, sample preparation is one of the most important steps.

Because metabolomics provides a biochemical fingerprint of the integrated response of an organism to internal or external stimuli, rapid quenching of all biochemical processes is the first important step. A rapid inactivation of metabolism is usually achieved through rapid changes in temperature or pH. Freezing in liquid nitrogen is the most reasonable way to stop enzymatic activity in plant and animal tissues. On the other hand, quenching a microbial cell culture is limited by the high dilution ratio between biomass and the extracellular medium. The use of precooled organic solvents such as cold methanol or cold ethanol has been reported as an excellent quenching method in eukaryotes (i.e., *Saccharomyces cerevisiae*). However, prokaryotic cells such as *Escherichia coli* tend to leak intracellular metabolites when exposed to the most universally used cold methanol protocol. Therefore, the rapid filtration method is still the preferable choice for prokaryotes quenching even for rapid sampling usage.

For analysis of intracellular metabolites, it is important to have a good extraction method that will extract all or the maximum number of cellular metabolites with the highest possible recoveries. There is no unique extraction procedure for metabolome analysis that will cover all intracellular metabolites. However, it is virtually impossible to avoid losses because of many different factors, particularly because of the large physicochemical properties of metabolites being quantified. The most preferable extraction method in the metabolomics area is modified Bligh and Dyer extraction (Bligh and Dyer, 1959) with methanol, chloroform, and water. On the other hand, in the medical fields, use of 80% methanol rather than Bligh and Dyer solvent for extraction is recommended because it is simpler and, therefore, likely to be more robust for high-throughput studies in particular. Therefore, the optimum extraction solvent should be developed depending on the purpose of the study.

References

Aharoni, A., Keizer, L. C. P., Bouwmeester, H. J., Sun, Z., Alvarez-Huerta, M., Verhoeven, H, A., Blaas, J., van Houwelingen, A. M. M. L., De Vos, R. C. H., and van der Voet, H., Identification of the SAAT gene involved in strawberry flavor biogenesis by use of DNA microarrays, *The Plant Cell Online*, 2000, 12: 647–661.

Al Bratty, M., Chintapalli, V. R., Dow, J. A. T., Zhang, T., and Watson, D. G., Metabolomic profiling reveals that *Drosophila melanogaster* larvae with the y mutation have altered lysine metabolism, *FEBS Open Biology*, 2012, 2: 217–221.

Arakaki, A. K., Skolnick, J., and McDonald, J. F., Marker metabolites can be therapeutic targets as well, *Nature*, 2008, 456: 443–443.

Bando, K., Kawahara, R., Kunimatsu, T., Sakai, J., Kimura, J., Funabashi, H., Seki, T., Bamba, T., and Fukusaki, E., Influences of biofluid sample collection and handling procedures on GC–MS based metabolomic studies, *Journal of Bioscience and Bioengineering*, 2010, 110: 491–499.

Beckonert, O., Keun, H. C., Ebbels, T. M. D., Bundy, J., Holmes, E., Lindon, J. C., and Nicholson, J. K., Metabolic profiling, metabolomic and metabonomic procedures for NMR spectroscopy of urine, plasma, serum and tissue extracts, *Nature Protocols*, 2007, 2: 2692–2703.

Bennett, B. D., Kimball, E. H., Gao, M., Osterhout, R., Van Dien, S. J., and Rabinowitz, J. D., Absolute metabolite concentrations and implied enzyme active site occupancy in *Escherichia coli*, *Nature Chemical Biology*, 2009, 5: 593–599.

Bennett, B. D., Yuan, J., Kimball, E. H., and Rabinowitz, J. D., Absolute quantitation of intracellular metabolite concentrations by an isotope ratio-based approach, *Nature Protocols*, 2008, 3: 1299–1311.

Blaise, B. J., Giacomotto, J., Elena, B., Dumas, M.-E., Toulhoat, P., Ségalat, L., and Emsley, L., Metabotyping of *Caenorhabditis elegans* reveals latent phenotypes, *Proceedings of the National Academy of Sciences*, 2007, 104: 19808–19812.

Bligh, E. G. and Dyer, W. J., A rapid method of total lipid extraction and purification, *Canadian Journal of Biochemistry and Physiology*, 1959, 37: 911–917.

Bosch, F. X., Ribes, J., Díaz, M., and Cléries, R., Primary liver cancer: Worldwide incidence and trends, *Gastroenterology*, 2004, 127: S5–S16.

Bubliy, O. A. and Loeschcke, V., Correlated responses to selection for stress resistance and longevity in a laboratory population of *Drosophila melanogaster*, *Journal of Evolutionary Biology*, 2005, 18: 789–803.

Cevallos-Cevallos, J. M. and Reyes-De-Corcuera, J. I., Metabolomics in food science, *Advances in Food and Nutrition Research*, 2011, 67: 1–24.

Cevallos-Cevallos, J. M., Reyes-De-Corcuera, J. I., Etxeberria, E., Danyluk, M. D., and Rodrick, G. E., Metabolomic analysis in food science: A review, *Trends in Food Science & Technology*, 2009, 20: 557–566.

Chintapalli, V. R., Al Bratty, M., Korzekwa, D., Watson, D. G., and Dow. J. A. T., Mapping an atlas of tissue-specific *Drosophila melanogaster* metabolomes by high resolution mass spectrometry, *PloS ONE,* 2013, 8: e78066.

Cook, K. R., Parks, A. L., Jacobus, L. M., Kaufman, T. C., and Matthews, K., New research resources at the Bloomington Drosophila Stock Center, *Fly*, 2010, 4: 88–91.

Crutchfield, C. A., Lu, W., Melamud, E., and Rabinowitz, J. D. Mass spectrometry-based metabolomics of yeast, *Methods in Enzymology,* 2010, 470: 393–426.

de Koning, W. and van Dam, K., A method for the determination of changes of glycolytic metabolites in yeast on a subsecond time scale using extraction at neutral pH, *Analytical Biochemistry*, 1992, 204: 118–123.

Dunn, W. B., Broadhurst, D., Begley, P., Zelena, E., Francis-McIntyre, S., Anderson, N., Brown, M., Knowles, J. D., Halsall, A., and Haselden, J. N., Procedures for large-scale metabolic profiling of serum and plasma using gas chromatography and liquid chromatography coupled to mass spectrometry, *Nature Protocols*, 2011, 6: 1060–1083.

Falk, M. J., Zhang, Z., Rosenjack, J. R., Nissim, I., Daikhin, E., Sedensky, M. M., Yudkoff, M., and Morgan, P. G., Metabolic pathway profiling of mitochondrial respiratory chain mutants in *C. elegans, Molecular Genetics and Metabolism,* 2008, 93: 388–397.

Fiehn, O., Metabolomics–The link between genotypes and phenotypes, *Plant Molecular Biology,* 2002, 48: 155–171.

Fiehn, O., Kopka, J., Dörmann, P., Altmann, T., Trethewey, R. N., and Willmitzer, L., Metabolite profiling for plant functional genomics, *Nature Biotechnology,* 2000a, 18: 1157–1161.

Fiehn, O., Kopka, J., Trethewey, R. N., and Willmitzer, L., Identification of uncommon plant metabolites based on calculation of elemental compositions using gas chromatography and quadrupole mass spectrometry, *Analytical Chemistry,* 2000b, 72: 3573–3580.

Geier, F. M., Want, E. J., Leroi, A. M., and Bundy, J. G., Cross-platform comparison of *Caenorhabditis elegans* tissue extraction strategies for comprehensive metabolome coverage, *Analytical Chemistry,* 2011, 83: 3730–3736.

Griffin, J. L. and Nicholls, A. W., Metabolomics as a functional genomic tool for understanding lipid dysfunction in diabetes, obesity and related disorders, *Pharmacogenomics,* 2006, 7: 1095–1107.

Guo, A. C., Jewison, T., Wilson, M., Liu, Y., Knox, C., Djoumbou, Y., Lo, P., Mandal, R., Krishnamurthy, R., and Wishart, D. S., ECMDB: The *E. coli* Metabolome Database, *Nucleic Acids Research,* 2013, 41: D625–D630.

Haruki, Y., Hiroaki, K., Jong-Chol, C., and Yasuo, O., Studies on polysaccharides from *Angelica acutiloba*—IV. Characterization of an anti-complementary arabinogalactan from the roots of *Angelica acutiloba* Kitagawa, *Molecular Immunology,* 1985, 22: 295–304.

Hayashi, S., Akiyama, S., Tamaru, Y., Takeda, Y., Fujiwara, T., Inoue, K., Kobayashi, A., Maegawa, S., and Fukusaki, E., A novel application of metabolomics in vertebrate development, *Biochemical and Biophysical Research Communications,* 2009, 386: 268–272.

Hirai, M. Y., Sugiyama, K., Sawada, Y., Tohge, T., Obayashi, T., Suzuki, A., Araki, R., Sakurai, N., Suzuki, H., and Aoki K., Omics-based identification of Arabidopsis Myb transcription factors regulating aliphatic glucosinolate biosynthesis, *Proceedings of the National Academy of Sciences,* 2007, 104: 6478–6483.

Hughes, S. L., Bundy, J. G., Want, E. J., Kille, P., and Stürzenbaum, S. R., The metabolomic responses of *Caenorhabditis elegans* to cadmium are largely independent of metallothionein status, but dominated by changes in cystathionine and phytochelatins, *Journal of Proteome Research*, 2009, 8: 3512–3519.

Hui, Y. H. and Evranuz, E. Ö., *Handbook of Plant-Based Fermented Food and Beverage Technology*, 2012, Boca Raton, FL: CRC Press.

Ibáñez, A. J., Fagerer, S. R., Schmidt, A. M., Urban, P. L., Jefimovs, K.., Geiger, P., Dechant, R., Heinemann, M., and Zenobi, R., Mass spectrometry-based metabolomics of single yeast cells, *Proceedings of the National Academy of Sciences*, 2013, 110: 8790–8794.

Ikeda, T., Kanaya, S., Yonetani, T., Kobayashi, A., and Fukusaki, E., Prediction of Japanese green tea ranking by Fourier transform near-infrared reflectance spectroscopy, *Journal of Agricultural and Food Chemistry*, 2007, 55: 9908–9912.

Intelmann, D., Haseleu, G., Dunkel, A., Lagemann, A., Stephan, A., and Hofmann, T., Comprehensive sensomics analysis of hop-derived bitter compounds during storage of beer, *Journal of Agricultural and Food Chemistry*, 2011, 59: 1939–1953.

Izrayelit, Y., Srinivasan, J., Campbell, S. L., Jo, Y., von Reuss, S. H., Genoff, M. C., Sternberg, P. W., and Schroeder, F. C., Targeted metabolomics reveals a male pheromone and sex-specific ascaroside biosynthesis in *Caenorhabditis elegans*, *ACS Chemical Biology*, 2012, 7: 1321–1325.

Jemal, A., Siegel, R., Xu, J., and Ward, E., 2010, Cancer statistics, *CA: A Cancer Journal for Clinicians*, 2010, 60: 277–300.

Jewison, T., Knox, C., Neveu, V., Djoumbou, Y., Guo, A. C., Lee, J., Liu, P., Mandal, R., Krishnamurthy, R., and Sinelnikov I., YMDB: The yeast metabolome database, *Nucleic Acids Research*, 2012, 40: D815–D820.

Jones, O. A. H., Swain, S., Svendsen, C. C., Griffin, J. L., Sturzenbaum, S. R., and Spurgeon, D. J., Potential new method of mixture effects testing using metabolomics and *Caenorhabditis elegans,* *Journal of Proteome Research*, 2012, 11: 1446–1453.

Jumhawan, U., Putri, S. P., Marwani, E., Bamba, T., and Fukusaki E., Selection of discriminant markers for authentication of Asian palm civet coffee (kopi luwak): A metabolomics approach, *Journal of Agricultural and Food Chemistry*, 2013, 61: 7994–8001.

Jumtee, K., Bamba, T., Okazawa, A., Fukusaki, E., and Kobayashi, A., Integrated metabolite and gene expression profiling revealing phytochrome A regulation of polyamine biosynthesis of *Arabidopsis thaliana*, *Journal of Experimental Botany*, 2008, 59: 1187–1200.

Kamath, R. S., Fraser, A. G., Dong, Y., Poulin, G., Durbin, R., Gotta, M., Kanapin, A., Le Bot, N., Moreno, S., and Sohrmann, M., Systematic functional analysis of the *Caenorhabditis elegans* genome using RNAi, 2003, *Nature*, 421: 231–237.

Kamleh, M. A., Hobani, Y., Dow, J. A. T., and Watson, D. G., Metabolomic profiling of *Drosophila* using liquid chromatography Fourier transform mass spectrometry, *FEBS Letters*, 2008, 582: 2916–2922.

Kamleh, M. A., Hobani, Y., Dow, J. A. T., Zheng, L., and Watson, D. G., Towards a platform for the metabonomic profiling of different strains of *Drosophila melanogaster* using liquid chromatography–Fourier transform mass spectrometry, *FEBS Journal*, 2009, 276: 6798–6809.

Kato, H., Izumi, Y.,Hasunuma, T., Matsuda, F., and Kondo, A., Widely targeted metabolic profiling analysis of yeast central metabolites, *Journal of Bioscience and Bioengineering*, 2012, 113: 665–673.

Kawase, N., Tsugawa, H., Bamba, T., and Fukusaki. E., Different-batch metabolome analysis of *Saccharomyces cerevisiae* based on gas chromatography/mass spectrometry, *Journal of Bioscience and Bioengineering,* 2014, 117(2): 248–255.

Kell, D. B., Brown, M., Davey, H. M., Dunn, W. B., Spasic, I., and Oliver, S. G., Metabolic footprinting and systems biology: The medium is the message, *Nature Reviews Microbiology*, 2005, 3: 557–565.

Khan, S. A., Chibon, P.-Y., de Vos, R. C. H., Schipper, B. A., Walraven, E., Beekwilder, J., van Dijk, T., Finkers, R., Visser, R. G. F., and van de Weg, E. W., Genetic analysis of metabolites in apple fruits indicates an mQTL hotspot for phenolic compounds on linkage group 16, *Journal of Experimental Botany*, 2012, 63: 2895–2908.

Ko, B.-K., Kim, K. M., Hong, Y.-S., and Lee, C.-H., Metabolomic assessment of fermentative capability of soybean starter treated with high pressure, *Journal of Agricultural and Food Chemistry*, 2010, 58: 8738–8747.

Lai, C.-H., Chou, C.-Y., Ch'ang, L.-Y., Liu, C.-S., and Lin, W.-C., Identification of novel human genes evolutionarily conserved in *Caenorhabditis elegans* by comparative proteomics, *Genome Research*, 2000, 10: 703–713.

Lee, J. W., Uchikata, T., Matsubara, A., Nakamura, T., Fukusaki, E., and Bamba T., Application of supercritical fluid chromatography/mass spectrometry to lipid profiling of soybean, *Journal of Bioscience and Bioengineering*, 2012, 113: 262–268.

Lei, Z., Huhman, D. V., and Sumner, L. W., Mass spectrometry strategies in metabolomics, *Journal of Biological Chemistry*, 2011, 286: 25435–25442.

Lu, Y., Lam, H., Pi, E., Zhan, Q., Tsai, S., Wang, C., Kwan, Y., and Ngai, S., Comparative metabolomics in *Glycine max* and *Glycine soja* under salt stress to reveal the phenotypes of their offspring, *Journal of Agricultural and Food Chemistry*, 2013, 61: 8711–8721.

Luo, Q., Yu, B., and Liu, Y., Differential sensitivity to chloride and sodium ions in seedlings of *Glycine max* and *G. soja* under NaCl stress, *Journal of Plant Physiology*, 2005, 162: 1003–1012.

Malmendal, A., Overgaard, J., Bundy, J. G., Sørensen, J. G., Nielsen, N. C., Loeschcke, V., and Holmstrup, M., Metabolomic profiling of heat stress: Hardening and recovery of homeostasis in *Drosophila*, *American Journal of Physiology-Regulatory, Integrative and Comparative Physiology*, 2006, 91: R205–R212.

Mashego, M. R., Rumbold, K., De Mey, M., Vandamme, E., Soetaert, W., and Heijnen, J. J., Microbial metabolomics: Past, present and future methodologies, *Biotechnology Letters*, 2007, 9:1–16.

Matsuda, F., Hirai, M. Y., Sasaki, E., Akiyama, K., Yonekura-Sakakibara, K., Provart, N. J., Sakurai, T., Shimada, Y., and Saito, K., AtMetExpress development: A phytochemical atlas of *Arabidopsis* development, *Plant Physiology*, 2010, 152: 566–578.

Matsuda, F., Yonekura-Sakakibara, K., Niida, R., Kuromori, T., Shinozaki, K., and Saito, K., MS/MS spectral tag-based annotation of non-targeted profile of plant secondary metabolites, *Plant Journal*, 2009, 57: 555–577.

Meinke, D. W., Cherry, J. M., Dean, C., Rounsley, S. D., and Koornneef, M., *Arabidopsis thaliana*: A model plant for genome analysis, *Science*, 1998, 282: 662–682.

Meyer, A., Biermann, C. H., and Orti, G., The phylogenetic position of the zebrafish (Danio rerio), a model system in developmental biology: An invitation to the comparative method, *Proceedings of the Royal Society of London. Series B: Biological Sciences*, 1993, 252: 231–236.

Neptun, D. A., Smith, C. N., and Irons, R. D., Effect of sampling site and collection method on variations in baseline clinical pathology parameters in Fischer-344 rats: 1. Clinical chemistry, *Fundamental and Applied Toxicology*, 1985, 5: 1180–1185.

Nishiumi, S., Kobayashi, T., Ikeda, A., Yoshie, T., Kibi, M., Izumi, Y., Okuno, T., Hayashi, N., Kawano, S., and Takenawa, T., A novel serum metabolomics-based diagnostic approach for colorectal cancer, *PloS ONE*, 2012, 7: e40459.

Nishiumi, S., Shinohara, M., Ikeda, A.,Yoshie, T., Hatano, N., Kakuyama, S., Mizuno, S., Sanuki, T., Kutsumi, H., and Fukusaki, E., Serum metabolomics as a novel diagnostic approach for pancreatic cancer, *Metabolomics*, 2010, 6: 518–528.

Ochi, H., Naito, H., Iwatsuki, K., Bamba, T., and Fukusaki, E., Metabolomics-based component profiling of hard and semi-hard natural cheeses with gas chromatography/time-of-flight-mass spectrometry, and its application to sensory predictive modeling, *Journal of Bioscience and Bioengineering*, 2012, 113: 751–758.

Ohashi, Y., Hirayama, A., Ishikawa, T., Nakamura, S., Shimizu, K.,Ueno, Y., Tomita, M., and Soga, T., Depiction of metabolome changes in histidine-starved *Escherichia coli* by CE-TOFMS, *Molecular BioSystems*, 2008, 4: 135–147.

Ong, E. S., Chor, C. F., Zou, L., and Ong, C. N., A multi-analytical approach for metabolomic profiling of zebrafish (*Danio rerio*) livers, *Molecular Biosystems*, 2009, 5: 288–298.

Organization, World Health, *International Classification of Diseases (ICD)*, 2013.

Pedersen, K. S., Kristensen, T. N., Loeschcke, V., Petersen, B. O., Duus, J. Ø., Nielsen, N. C., and Malmendal, A., Metabolomic signatures of inbreeding at benign and stressful temperatures in *Drosophila melanogaster*, *Genetics*, 2008, 180: 1233–1243.

Piao, X.-L., Park, J. H.,Cui, J., Kim, D.-H., and Yoo, H. H., Development of gas chromatographic/mass spectrometry-pattern recognition method for the quality control of Korean *Angelica*, *Journal of Pharmaceutical and Biomedical Analysis*, 2007, 44: 1163–1167.

Pongsuwan, W., Bamba, T., Harada, Yonetani, K. T., Kobayashi, A., and Fukusaki, E., High-throughput technique for comprehensive analysis of Japanese green tea quality assessment using ultra-performance liquid chromatography with time-of-flight mass spectrometry (UPLC/TOF MS), *Journal of Agricultural and Food Chemistry*, 2008, 56: 10705–10708.

Pongsuwan, W., Fukusaki, E., Bamba, T., Yonetani, T.,Yamahara, T., and Kobayashi, A., Prediction of Japanese green tea ranking by gas chromatography/mass spectrometry-based hydrophilic metabolite fingerprinting, *Journal of Agricultural and Food Chemistry*, 2007, 55: 231–236.

Putluri, N., Shojaie, A., Vasu, V. T.,Vareed, S. K., Nalluri, S., Putluri, V., Thangjam, G. S., Panzitt, K., Tallman, C. T., and Butler, C., Metabolomic profiling reveals potential markers and bioprocesses altered in bladder cancer progression, *Cancer Research*, 2011, 71: 7376–7386.

Putri, S. P., Bamba, T., and Fukusaki. E., Application of metabolomics for discrimination and sensory predictive modeling of food products, in *Hot Topics in Metabolics: Food and Nutrition*, Future Science, London, UK, 2013, 54–64.

Putri, S. P., Nakayama, Y., Matsuda, F., Uchikata, T., Kobayashi, S., Matsubara, A., and Fukusaki, E., Current metabolomics: Practical applications, *Journal of Bioscience and Bioengineering*, 2013, 115: 579–589.

Rabinowitz, J. D. and Kimball, E., Acidic acetonitrile for cellular metabolome extraction from *Escherichia coli, Analytical Chemistry*, 2007, 79: 6167–6173.

Rhee, S. Y., Beavis, W., Berardini, T. Z., Chen, G., Dixon, D., Doyle, A., Garcia-Hernandez, M., Huala, E., Lander, G., and Montoya, M., The *Arabidopsis* Information Resource (TAIR): A model organism database providing a centralized, curated gateway to *Arabidopsis* biology, research materials and community, *Nucleic Acids Research*, 2003, 31: 224–228.

Rodrigues, J. A., Barros, A. S., Carvalho, B., Brandão, T., and Gil, A. M., Probing beer aging chemistry by nuclear magnetic resonance and multivariate analysis, *Analytica Chimica Acta*, 2011, 702: 178–187.

Soanes, K. H., Achenbach, J. C., Burton, I. W., Hui, J. P. M., Penny, S. L., and Karakach, T. K., Molecular characterization of zebrafish embryogenesis via DNA microarrays and multiplatform time course metabolomics studies, *Journal of Proteome Research*, 2011, 10: 5102–5117.

Somerville, C. and Koornneef, M., A fortunate choice: The history of *Arabidopsis* as a model plant, *Nature Reviews Genetics*, 2002, 3: 883–889.

Son, H.-S., Kim, K. M.,Van Den Berg, F., Hwang, G.-S., Park, W.-M., Lee, C.-H., and Hong, Y.-S., ^{1}H nuclear magnetic resonance-based metabolomic characterization of wines by grape varieties and production areas, *Journal of Agricultural and Food Chemistry*, 2008, 56: 8007–8016.

Sreekumar, A., Poisson, L. M., Rajendiran, T. M., Khan, A. P., Cao, Q., Yu, J., Laxman, B., Mehra, R., Lonigro, R. J., and Li, Y., Metabolomic profiles delineate potential role for sarcosine in prostate cancer progression, *Nature*, 2009, 457: 910–914.

Stern, H. M. and Zon, L. I., Cancer genetics and drug discovery in the zebrafish, *Nature Reviews Cancer*, 2003, 3: 533–539.

Strange, K., Christensen, M., and Morrison, R., Primary culture of *Caenorhabditis elegans* developing embryo cells for electrophysiological, cell biological and molecular studies, *Nature Protocols*, 2007, 2: 1003–1012.

Sugimoto, M., Ikeda, S., Niigata, K., Tomita, M., Sato, H., and Soga, T., MMMDB: Mouse multiple tissue metabolome database, *Nucleic Acids Research*, 2012, 40: D809–D814.

Sugimoto, M., Koseki, T., Hirayama, A., Abe, S., Sano, T., Tomita, M., and Soga, T., Correlation between sensory evaluation scores of Japanese sake and metabolome profiles, *Journal of Agricultural and Food Chemistry*, 2009, 58: 374–383.

Tarachiwin, L., Katoh, A., Ute, K., and Fukusaki, E., Quality evaluation of *Angelica acutiloba* Kitagawa roots by ^{1}H NMR-based metabolic fingerprinting, *Journal of Pharmaceutical and Biomedical Analysis*, 2008, 48: 42–48.

Tarachiwin, L., Ute, K., Kobayashi, A., and Fukusaki, E., ^{1}H NMR based metabolic profiling in the evaluation of Japanese green tea quality, *Journal of Agricultural and Food Chemistry*, 2007, 55: 9330–9336.

Taymaz-Nikerel, H., De Mey, M., Ras, C., ten Pierick, A., Seifar, R. M., Van Dam, J. C., Heijnen, J. J., and van Gulik, W. M., Development and application of a differential method for reliable metabolome analysis in *Escherichia coli*, *Analytical Biochemistry*, 2009, 386: 9–19.

Tianniam, S., Bamba, T., and Fukusaki, E., Pyrolysis GC-MS-based metabolite fingerprinting for quality evaluation of commercial *Angelica acutiloba* roots, *Journal of Bioscience and Bioengineering*, 2010, 109 (1): 89–93.

Tianniam, S., Tarachiwin, L., Bamba, T., Kobayashi, A., and Fukusaki, E., Metabolic profiling of *Angelica acutiloba* roots utilizing gas chromatography–time-of-flight–mass spectrometry for quality assessment based on cultivation area and cultivar via multivariate pattern recognition, *Journal of Bioscience and Bioengineering*, 2008, 105: 655–659.

Vaclavik, L., Lacina, O., Hajslova, J., and Zweigenbaum, J., The use of high performance liquid chromatography–quadrupole time-of-flight mass spectrometry coupled to advanced data mining and chemometric tools for discrimination and classification of red wines according to their variety, *Analytica Chimica Acta*, 2011, T 685: 45–51.

Varga, Z. M., Aquaculture and husbandry at the zebrafish international resource center, *Methods in Cell Biology*, 2011, 104: 453.

Villas-Bôas, S. G., Højer-Pedersen, J., Åkesson M., Smedsgaard, J., and Nielsen, J., Global metabolite analysis of yeast: Evaluation of sample preparation methods, *Yeast*, 2005, 22: 1155–1169.

von Reuss, S. H., Bose, Srinivasan, N. J., Yim, J. J., Judkins, J. C., Sternberg, P. W., and Schroeder, F. C., Comparative metabolomics reveals biogenesis of ascarosides, a modular library of small-molecule signals in *C. elegans*, *Journal of the American Chemical Society*, 2012, 134: 1817–1824.

Weckwerth, W., Loureiro M. E., Wenzel, K., and Fiehn, O., Differential metabolic networks unravel the effects of silent plant phenotypes, *Proceedings of the National Academy of Sciences of the United States of America*, 2004a, 101: 7809–7814.

Weckwerth, W., Wenzel, K., and Fiehn, O., Process for the integrated extraction, identification and quantification of metabolites, proteins and RNA to reveal their co-regulation in biochemical networks, *Proteomics*, 2004b, 4: 78–83.

Winder, C. L., Dunn, W. B., Schuler, S., Broadhurst, D., Jarvis, R., Stephens, G. M., and Goodacre R., Global metabolic profiling of *Escherichia coli* cultures: An evaluation of methods for quenching and extraction of intracellular metabolites, *Analytical Chemistry*, 2008, 80: 2939–2948.

Wishart, D. S., Tzur, D., Knox, C., Eisner, R., Guo, A. C., Young, N., Cheng, D., Jewell, K., Arndt, D., and Sawhney, S., HMDB: The human metabolome database, *Nucleic Acids Research*, 2007, 35: D521–D526.

Wu, H., Xue, R., Dong, L., Liu, T., Deng, C., Zeng, H., and Shen, X., Metabolomic profiling of human urine in hepatocellular carcinoma patients using gas chromatography/mass spectrometry, *Analytica Chimica Acta*, 2009, 648: 98–104.

Yamamoto, S., Bamba, T., Sano, A., Kodama, Y., Imamura, M., Obata, A., and Fukusaki, E., Metabolite profiling of soy sauce using gas chromatography with time-of-flight mass spectrometry and analysis of correlation with quantitative descriptive analysis, *Journal of Bioscience and Bioengineering*, 2012, 114: 170–175.

Yuan, J., Doucette, C. D., Fowler, W. U., Feng, X.-J., Piazza, M., Rabitz, H. A., Wingreen, N. S., and Rabinowitz, J. D., Metabolomics-driven quantitative analysis of ammonia assimilation in *E. coli*, *Molecular Systems Biology*, 2009, 5: 302.

4

Gas Chromatography/ Mass Spectrometry Analysis

Nontargeted Metabolomics Based on Scan Mode Analysis

Chapter 4

Gas Chromatography/ Mass Spectrometry Analysis

Hiroshi Tsugawa and Arjen Lommen

Chapter Outline

4.1 Introduction

Metabolome analysis is divided into two approaches: nontargeted analysis and targeted analysis.[1] Nontargeted analysis covers all detected peaks by means of scan mode and utilizes both annotated and unannotated peak information for statistical analyses. This approach has been broadly applied to discovery-phase studies such as biomarker identification for medical diagnostics. On the other hand, the targeted or widely targeted analysis hunts only "known metabolites" detected by single ion monitoring or selected reaction monitoring mode with the purpose of validating a biological hypothesis or machine-learning studies such as regression and discriminant analyses. In this chapter, we introduce the protocol for the nontargeted gas chromatography/mass spectrometry (GC/MS) based metabolomics.

GC/MS is one of the most frequently used platforms in metabolomics study for comprehensive analysis of low-molecular–weight primary metabolites because of its low cost, stability, reproducibility, high peak capacity, and relatively easy data processing.[2] It usually provides information on several hundreds of metabolites, which are easily identified owing to their highly reproducible retention time and mass spectra. Scan mode analysis is usually utilized for data acquisition of mass spectra in GC/MS for nontargeted analysis. Ever since the report of the identification of 164 metabolites in *Arabidopsis thaliana* rosette leaf by nontargeted GC/MS analysis to clarify silent gene characters by metabolome phenotype successfully, GC/MS has been realized as a core technology in metabolomics and has been applied to many research fields such as food science and biomarker discovery.[3–6]

Nontargeted GC/MS metabolomics mainly consists of five steps, namely: sample extraction, MEOX-TMS-derivatization, GC/MS analysis by scan mode, data processing, and statistical analyses. This chapter provides an easy-to-use protocol for nontargeted GC/MS analysis that is recommended as a startup of metabolomics.

4.2 Advantages and Disadvantages of GC/MS Analysis

The advantages of GC/MS are the stability of the platform, high reproducibility of retention time, and high reproducibility of mass spectra. Because GC/MS is a mature technology applicable to a large number of samples, researchers can easily perform the instrument operation and maintenance. Moreover, after calculating the retention time index from *n*-alkane or fatty acid methyl ester, the detected peaks can be easily compared to some databases such as the Fiehn library,[7] MassBank,[8] and Golm Metabolome Database.[9] In addition, as much as mass fragmentation from a metabolite is produced by the electron ionization method standardized to 70 eV, its mass spectra information is compatible with any instrument or laboratory. This indicates that researchers can utilize a freely available and reliable reference library to perform nontargeted GC/MS metabolomics easily without preparing the reference library themselves. However, in order to identify the metabolites accurately (this is the most important issue), the user should follow completely the same analytical settings that were used to construct the reference library because the retention index would be adjusted. The retention "time" is not recommended as the index for peak identification because the retention index would be slightly different even if the same analytical conditions were used.

The disadvantage of GC/MS is the need for a derivatization process. Methoxyamine (MEOX) derivatization is used for capping of highly polar carbonyl groups and trimethylsilyl (TMS) derivatization is used for capping of highly polar hydrogen groups such as alcohol, carboxylic acid, and amine group. These derivatization processes contribute to the improvement of peak tailing and reduction of the boiling point for highly volatile compounds, however, they also encompass some undesirable situations for metabolite identification and quantification.[10–12] For example, the amino acid group mostly generates two peaks due to different degrees of silylation at the primary or secondary amines resulting from a poorly reactive and unstable TMS derivatization. Moreover, the sugar group mostly generates several peaks due to their geometric isomers derived from the oxime reaction. Such peaks prevent the accurate identification and quantification of metabolites. Some metabolites might also be generated by additional reactions such as the pyrolysis reaction in the front inlet and capillary column. The formation of TMS-pyroglutamate from TMS-glutamate is a characteristic example of an additional reaction from the derivatization process.[13]

Moreover, the mono-phosphate peak is always generated from sugar phosphate or nucleotide. Thus, these features indicate that there are some "black boxes" in the GC/MS analysis and researchers have to recognize this biological aspect when they discuss the metabolome data using GC/MS-based metabolomics.

4.3 Analytical Method of GC/MS Analysis

GC/MS is one of the most frequently used platforms for non-targeted analysis. In this method, hydrophilic low-molecular–weight metabolites such as sugar, amino acid, and organic acid are derivatized by methoxymation and silylation. There are many coeluted metabolites and isomeric metabolites with mostly the same retention time and mass spectra, therefore data processing can be time consuming and expert knowledge for peak identification is required. To overcome this problem, Fukusaki and colleagues have developed a data processing system and freely available metabolite library.[13] Here, we introduce an easy-to-use workflow using a Shimadzu GC/MS instrument, MetAlign[14,15] for peak detection and alignment, and finally AIoutput for peak identification and data integration. In the case where another instrument is used for GC/MS analysis, the same analytical conditions are recommended to get a meaningful result.

4.3.1 Theory

We modified the method used in the Fiehn laboratory with consideration of the running cost.[16,17] In our method, the sample is extracted using a mixed solvent of MeOH, H_2O, and $CHCl_3$ and the internal standard (ribitol, adipic acid, or 2-isopropylmalic acid) is also added into the mixed solvent at the same time. The details of sample extraction are explained in Chapter 3.

4.3.2 Materials and Instrument Setting

4.3.2.1 Reagents

- Pyridine (Wako, Infinity Pure 99.8+%, Catalog No. 169-17011)

- Methoxyamine hydrochloride (Sigma Aldrich, Catalog No. 226904)
- *N*-Methyl-*N*-trimethylsilyltrifluoroacetamide (MSTFA) (GL Sciences, 1g in ampoule glass tube, Catalog No. 1022-11061)
- Ribitol (Wako, Catalog No. 018-00943)
- Alkane standard solution C8-C20 (Sigma Aldrich, Catalog No. 4070, 5 mL)
- Alkane standard solution C21-C40 (Sigma Aldrich, Catalog No. 4071, 5 mL)

4.3.2.2 *Apparatuses*

- Column (30 m × 0.25 mm i.d. fused silica capillary column coated with 0.25-µm CP-SIL 8 CB low bleed, GL Science, Catalog No. 1010-78751)
- Vial (Chromacol Crimp Top Vial, GL Science, Catalog No. 1030-41360)
- Crimp cap (Chromacol, GL Science, Catalog No. 1030-42425)
- 2.0-mL micro-tube (Eppendorf, Safe lock tube, Catalog No. 95170)
- Glass insert (RESTEK Catalog No. 20956, Split/Splitless Liner, 3.5 mm × 5.0 × 95 for Shimadzu GCs)

4.3.2.3 *Instruments*

- Autosampler: AOC-20is series injector (Shimadzu Inc.)
- Gas chromatograph: GC-2010 Plus (Shimadzu Inc.)
- Mass spectrometer: GCMS-QP2010 Ultra (Shimadzu Inc.)
- Incubator (Thermomixer comfort, Eppendorf)
- Vortex mixer

4.3.2.4 *GC/MS Conditions*

- Inlet temperature: 230°C
- Injection volume: 1 µL

- Injection mode: Split
- Split ratio: 25.0
- Oven temperature program: 80°C, 2-min hold, 15°C/min-gradient, 330°C, 6-min hold
- Gas flow rate: Helium 1.12 mL/min
- Purge flow rate: 5.0 mL/min
- Interface temperature: 250°C
- Ion source temperature: 200°C
- Ionization voltage: 70 eV
- Analysis mode: Scan
- Scan range: m/z 85–500
- Scan speed: 10,000 u/s
- Gain voltage: 0.9 kV (adjustment in a batch analysis)
- Data acquisition delay: 3 min
- Data acquisition period: 3.5 min–24.0 min

4.3.2.5 Alkane Mixture

- The user should analyze the n-alkane mixture in order to calculate the retention index and use our reference library.
- Mix 50 µL of alkane standard solution C8-C20, 50 µL of alkane standard solution C21-C40, and 50 µL of pyridine in a vial.

4.3.2.6 Quality Control (QC) Sample

- Quality control sample should also be prepared to check the sensitivity and correct the quantification value of metabolites.
- It is recommended to mix small aliquots of each biological sample (e.g., yeast extract). The QC sample should be processed by the same protocol used for each sample.
- The quantitative correction method by means of the QC sample has been reported.[18] This is highly important, particularly for the data integration among the experimental batches in a large-scale experiment.

4.3.3 Experimental Protocol

Sample derivatization (for Japanese green tea):

1. Add 100 µL of methoxyamine hydrochloride in pyridine (20 mg/mL) to a freeze-dried sample (for yeast: 50 µL, for serum or plasma: 80 µL).
2. Incubate at 30°C, 1,200 rpm for 90 min.
3. Add 50 mL of N-methyl-N-trimethysilyltrifluoroacetamide (MSTFA) to the sample (for yeast: 25 µL, for serum or plasma: 40 µL).
4. Incubate at 37°C, 1,200 rpm for 30 min.
5. Centrifuge at 4°C, 16,000 rpm for 5 min.
6. Transfer the supernatant to a vial.

Note:: The sample and all apparatuses should be completely dried to prevent degradation of derivatized metabolite by hydrolysis reaction. If sonication process is required to dissolve a dried sample, keep at low temperature by using ice to prevent heat degradation of metabolites.

Samples should be analyzed up to 24 hours after derivatization because the TMS-metabolite is easily degraded.

A blank sample should also be prepared to check for contamination.

4.3.4 GC/MS-QP2010 Ultra Operation

(1) In the "Ecology Mode", push "Cancel" button (Figure 4.1).
(2) Leak check:
  a. Check the leakage monitor visually. Then, "Save current Tuning file?" = > No.
  b. Click "Tuning" = > "Peak monitor view" of the assistant toolbar (Figure 4.2).
① Select "H_2O, N_2" in "Monitor Group".
② Click the filament icon. = > Filament ON.
③ Adjust the "Detector" voltage until the peak height of m/z 18 (H_2O) reaches half of the height of the displayed window of GC/MS solution software as described in the middle of Figure 4.2.

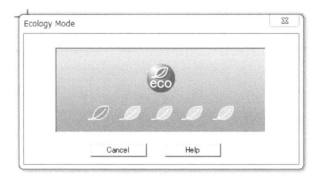

Figure 4.1 Screen shot of ecology mode. (Reprinted with permission from Shimadzu Inc.)

④ Confirm that the peak height of m/z 28 is not more than twice that of m/z 18.

⑤ Print out the leak check result.

⑥ Click the filament icon. = > Filament OFF.

(3) Create a new save folder file.

    a. "Data acquisition" = > "Data explorer of the Assistant Bar" = > "New folder" = > Define the file name (Ex. Operator name_yyyymmdd).

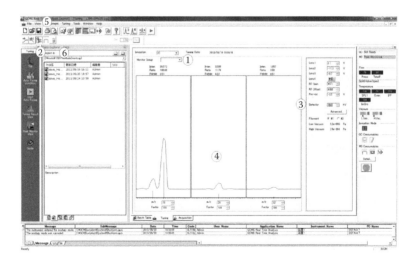

Figure 4.2 Screen shot of peak monitor view. (Reprinted with permission from Shimadzu Inc.)

(4) Create a method file.
   a. Assistant toolbar "Data acquisition" = > Create or copy a method file to your own folder (Figures 4.3 and 4.4.

Note: The method used in the Fukusaki laboratory is described below.

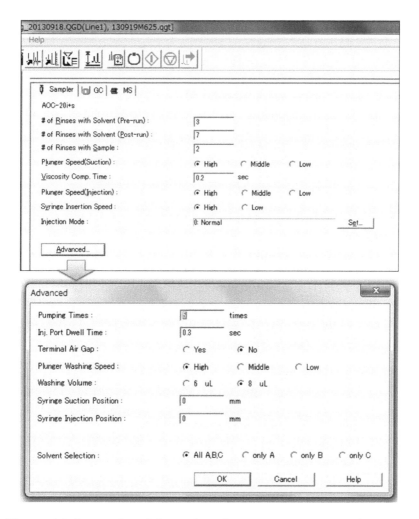

Figure 4.3 Screen shot of the analytical parameter setting window of gas chromatography. (Reprinted with permission from Shimadzu Inc.)

To apply the acquired data set to the AIoutput identification data processing system, the scan range must be set from m/z 85 to m/z 500.

In GC/MS-based metabolomics, the same method is always used thus the same method file can be copied and shared once a method file is created.

The analytical parameter setting window of gas chromatography and mass spectrometry can be seen in Figures 4.3 and 4.4, respectively.

(5) Auto tuning: Tuning should be done for every experiment.
   a. Load or create a method file.
   b. "Data acquisition" => "Download the initial value of Acquisition" => Run auto tuning. Tuning is carried out using the instrument condition of your method file.
   c. Print out and check the result as follows:

- All of the full width at half maximum between 0.5 and 0.7 m/z.
- The detector voltage should be less than 2 kV.
- The base peak should be m/z 69 or m/z 18.
- The relative abundance of m/z 502 should be less than 2%.
- The amplitude of m/z 69 should be more than twice the amplitude of m/z 28.

(6) Syringe setting.
   a. Clean the syringe using acetone, methanol, or hexane. The plunger should also be cleaned because the derivative reagent is highly reactive.
   b. Fix the syringe in the auto sampler. See the GC/MS solution manual.
   c. Pyridine should be used as the cleaning reagent for the auto sampler.

(7) Create a batch file and analyze: The following sequence is recommended:

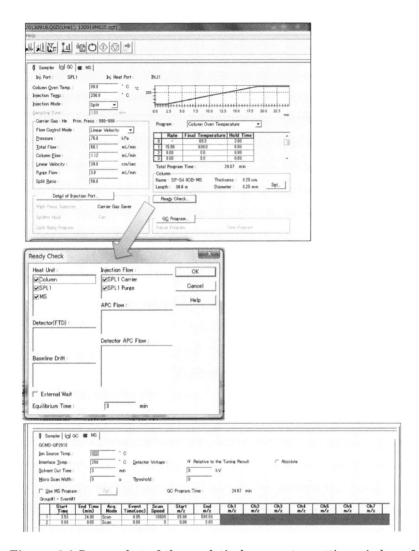

Figure 4.4 Screen shot of the analytical parameter setting window of mass spectrometry. (Reprinted with permission from Shimadzu Inc.)

a. Pyridine (to check the background)
b. Blank sample (to check the contamination)
c. Alkane mixture (to get the retention time of *n*-alkane for retention index calculation)
d. Pyridine (to clean up the carryover of *n*-alkane)
e. Pyridine (to clean up the carryover of *n*-alkane)

f. After a–e injections, the biological samples are analyzed. Make sure that 1-QC sample is also analyzed for every 5–7 biological samples.

g. Finally, the blank sample and pyridine should be analyzed to check for contamination and to clean up the column.

Note: To change a batch file: "Batch processing" = > "Pause/Restart" = > Edit your batch file = > "Save Batch file as" = > "Pause/Restart"

(8) Final.
a. Set the "Eco mode". Because you will be asked whether the eco mode is set or not before you start a batch analysis, select "Yes".

4.3.5 Maintenance

Here are some important notes to consider when performing this experiment. Also, it is strongly recommended to carefully read the user manual of the MS vendor.

- Helium gas exchange: When the pressure is less than 5.0 MPa, a reserve should be prepared. If the pressure is less than 4.0 MPa, the exchange should be replaced with a new one.

- Septum, liner exchange: Replace both septum and liner if the number of total injections is beyond 100. After fastening the nut, slightly loosen the nut by turning it 90–180° in the opposite direction. Make sure that the nut is not tightly fastened to prevent errors in syringe insertion and septum deterioration. In cases where the O-ring is tightly adhered to the inlet entrance, tweezers should be used with caution when picking up the O-ring. There is a rope for replacing the new liner. When fixing the new liner, joint by the O-ring, temporarily. Then, ratchet down the liner until the top of insert is in contact with the bottom.

- Column exchange: Replace the column after 1,500 injections. This process will take several hours because it is essential to shut down the MS prior to column exchange. When the column apex is cut by a capillary cutter to connect it to both the front inlet and MS, confirm its straightness by using a magnifying glass. Because the most common cause of leakage comes from the contact site (especially the O-ring), carefully clean up the O-ring by acetone, methanol, or toluene. Finally, check the leakage using a leak detector.

- Others: Clean up the ion source and replace the filament every month. Replace the syringe if broken or contaminated. Replace the oil of the rotary pump every 6 months.

4.3.6 Troubleshooting

Some well-known problems of GC/MS analysis are saturated or undetected metabolites and carryover. These problems are mostly derived from the fact that the sample concentration is not optimized. In order to get high-quality data, it is important to validate the sample volume for GC/MS analysis. Based on the criteria below, the split ratio or sample volume should be set.

- Prepare the diluted samples and check the number of detected peaks and their amplitudes.
- The highest amplitude should be adjusted to 2–30,000,000 cps.
- If the highest amplitude is larger than the above criteria, dilute the sample.
- Other: (1) Check the peak symmetry. (2) Check the peak saturation. (3) Analyze a pyridine after analyzing the sample. Then, check for carryover.

4.4 Data Processing of GC/MS

In GC/MS-based metabolomics peak detection, chromatogram alignment and identification are critical data-processing steps in order to get an organized data matrix (e.g., for statistical analysis). Fukusaki and colleagues developed an automated data-processing system for a large number of GC/MS data sets and a reference library including the retention time and mass spectra of 475 metabolites.[13] As long as the same analytical measurement conditions are used, the researcher can automatically get an organized data matrix that includes 100–200 identified metabolites as well as 100–200 unknown metabolites. The peak detection and peak alignment are carried out by freely available software called MetAlign.[14,15] Component detection and identification from MetAlign results are carried out by freely available software, AIoutput. Here, we introduce the data analysis protocol.

4.4.1 Software

- MetAlign: http://www.wageningenur.nl/en/show/ MetAlign-1.htm
- RedMSViewer: https://dl.dropbox.com/u/75158272/ viewer.zip
- AIoutput: http://prime.psc.riken.jp/(Please follow the link: "Targeted and Non-targeted analysis software" = > "AIoutput")

4.4.2 Environment

- Windows OS (Recommended: Core i5 or more, RAM 4G or more)
- Microsoft Excel 2007 or later

(1) File conversion
   Convert the raw data of your MS vendor to netCDF data format. In the case of Shimadzu, follow the steps below:
   a. In the "Data Explorer", select raw data files.
   b. After file selection (files highlighted), right click = > File Convert = > To AIA (Andi) File
   c. Select an export folder = > OK

Note: AIA is the abbreviation for the Analytical Instrument Association. This institution defined the AIA format and its extension (.CDF). AIA is also sometimes called Andi (Analytical Data Interchange).

The required time for file conversion is more than one minute per file although the total time of conversion is dependent on the machine specifications.

(2) MetAlign data processing
    a. Download metAlign at www.metalign.nl. Here, the latest version 041012 is explained.

    b. Installation of metAlign. Right click "setup.exe" = > Run as administrator = > Click "Complete installation of metAlign" (It takes 1 second to register some dll's and ocx files; Figure 4.5.).

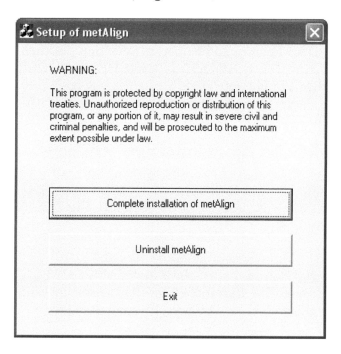

Figure 4.5 Screen shot of the setup window of metAlign. (Reprinted with permission from Dr. Arjen Lommen.)

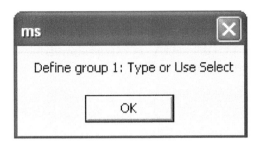

Figure 4.6 Screen shot of the setup message box from metAlign. (Reprinted with permission from Dr. Arjen Lommen.)

c. Open "ms.exe". First time use requires running it as administrator. You will get two start-up messages as shown below and OK should be clicked (Figure 4.6).

d. 1A. Program configuration (Figure 4.7).

① Use the default settings for "Definition of Folders".

② In "Data Format and Function Selection", confirm that the "INPUT FORMAT" is "NetCDF" and "OUTPUT FORMAT" is "NetCDF".

③ For users using multicore PCs such as Core i5 or i7 with a Windows 7, 64-bit operating system, the number of processors that can be used may be set 2, 3, or 4 (depending on the available RAM; 2-Gb RAM per processor is needed). For 32-bit systems only one processor should be used to avoid memory problems (32-bit systems can only address 2 Gb of RAM). Click the "Save" button.

e. 1B. Mass Resolution/Bin (Figure 4.8).

① Check "Nominal mass data" in SELECT DATA TYPE.

② In "Mass Bin Parameter for Conversion to Nominal", the suitable value is 0.6 for Shimadzu, 0.65 for LECO, and 0.7 for Agilent Mass spectrometer.

③ Click the OK button.

f. 2B. Select.

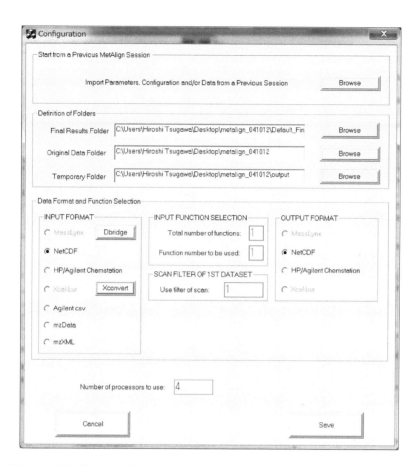

Figure 4.7 Screen shot of the program configuration of metAlign. (Reprinted with permission from Dr. Arjen Lommen.)

① Select the data files for processing.
② If you have previously selected a file, first click "Clear" button next to 2B to delete the reference of previous file paths.
③ 2A button is for checking/viewing the selected files. Manual changes are possible but not recommended (to avoid errors). The first file in the list has a special function to serve as a master or reference file for the alignment of all other files. It is therefore important

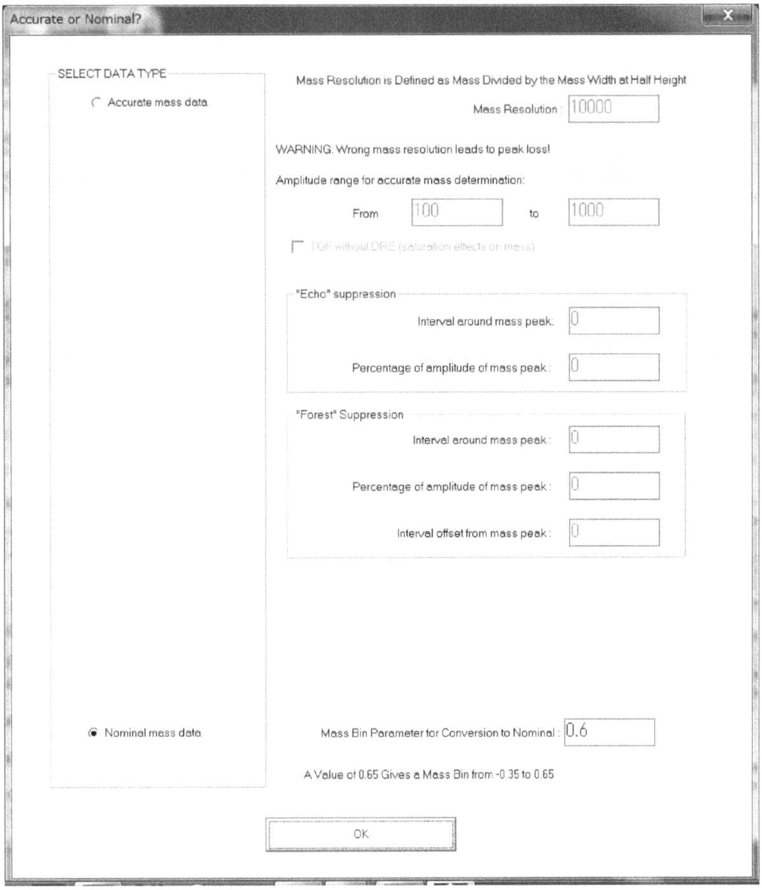

Figure 4.8 Screen shot of the mass bin setting window of metAlign. (Reprinted with permission from Dr. Arjen Lommen.)

to choose this well to obtain good alignment results in PART B. It is recommended to select the "middle" chromatogram (a file closest to the average retention profile) as the master/reference file (see below: choose by using the RedMSviewer). The users don't have to use 3A, 3B here.

g. The parameter setting of Part A (Figure 4.9)

① A detailed description of each parameter is given in the manual of metAlign. Here, only the suitable parameters for the above experimental conditions are shown.

Figure 4.9 Screen shot of Part A: data processing parameters for peak detection of metAlign. (Reprinted with permission from Dr. Arjen Lommen.)

h. Click 11. "Run Baseline Correction".

i. RedMSViewer.

In order to decide which file to use as master/reference and the first in the data list, the RedMSViewer (recently introduced in *MetaboNEWS,* 16 December, 2012) is very useful. It can also be used to determine suitable alignment parameters for part B. At first, download the RedMSViewer archive and extract. Open RedMSViewer.htm. This program runs locally on a web browser. Here, we describe how to use the RedMSViewer. (Allow blocked content; Silverlight needs to be installed first; Figure 4.10).

① Data import: Select all.redms files in the "Baseline" folder of your final results folder. The user can see all preprocessed files that are baseline-corrected and size-decreased (–hundredfold) chromatograms. By clicking and dragging the left button of the mouse, the chromatograms are simultaneously zoomed. This way, users can examine the preprocessed results (selected peaks) of all chromatograms. By clicking the left button of the mouse at the top area in each chromatogram picture box, the user can take the chromatogram and rearrange their order.

Figure 4.10 Tool buttons of RedMSViewer. (Reprinted with permission from Dr. Arjen Lommen.)

⑧ Change settings: Set the horizontal axis to "Retention" (by option 8 or double-click on the horizontal axis); this option is best for overlaying all data files.

② Reset the range of all graphics.

③ Go to the previous range of all graphics.

④ Create a new chromatogram overview of all superimposed files. Check "inverse selection" and "create a single chromatogram containing all selected datasets". Set the mass range from 319 to 319; this mass is a characteristic of the internal standard (ribitol) peak, which is present in all data files. From this overview, choose a signal for the master/reference data file (see above). By remaking new chromatograms but unchecking "create a single chromatogram containing all selected datasets" obtain all mass 319 traces of the data files and look for the name of the file with the chosen signal. This is defined as the first data file in 2A prior to alignment in PART 2B (Figure 4.11).

⑤ Create a heat map: this is also useful for checking the master/reference chromatogram and the max shift parameter (13. INITIAL PEAK SEARCH CRITERIA) needed as a measure for chromatogram displacement (retention shift; Figure 4.12).

⑥ Download your updated files. (Obsolete because we work offline).

⑦ Upload your analysis files. (Obsolete because we work offline.)

    j. Set the master/reference file (first file on the list).

① As long as the same analytical conditions are used, the user doesn't have to change parameters 4, 5, 6, 7, 8A, 8B, 9, 12, 14, 15, 16, 17, and 18. However, the alignment target file and the max shift (defined in parameter 13) should be changed depending on the GC-MS series.

② Using the RedMSViewer, check the "central" chromatogram file that stands in the middle position of all chromatograms.

③ Click "2A. Group 1: List of Data sets" button, to see the file path list that was selected before. As shown below, if "No21_1.cdf" file is used as the alignment target file, it

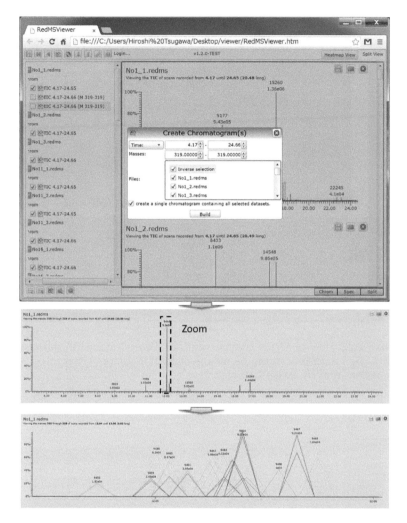

Figure 4.11 Selection method of the master data file by means of RedMSViewer. (Reprinted with permission from Dr. Arjen Lommen.)

should be taken to the top row by the cut and paste function. Alternatively, reselect the data using 2B and make sure you select the master file first and then the rest.
④ Save the rearranged text file (Figure 4.13).

k. The parameter setting of Part B (Figure 4.14).
① The details on each parameter are given in the manual of metAlign. Here, the suitable parameters for the

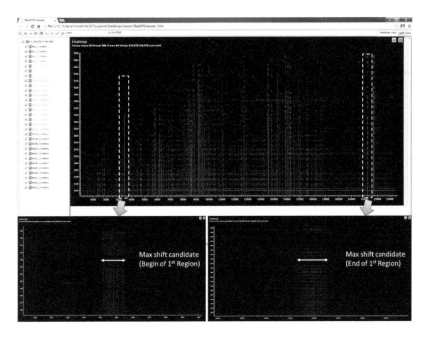

Figure 4.12 How to decide the master data file and max shift parameter in metAlign data processing by means of RedMSViewer. (Reprinted with permission from Dr. Arjen Lommen.)

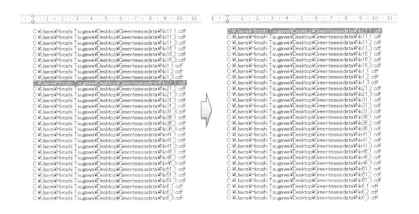

Figure 4.13 How to change the master data file in metAlign. (Reprinted with permission from Dr. Arjen Lommen.)

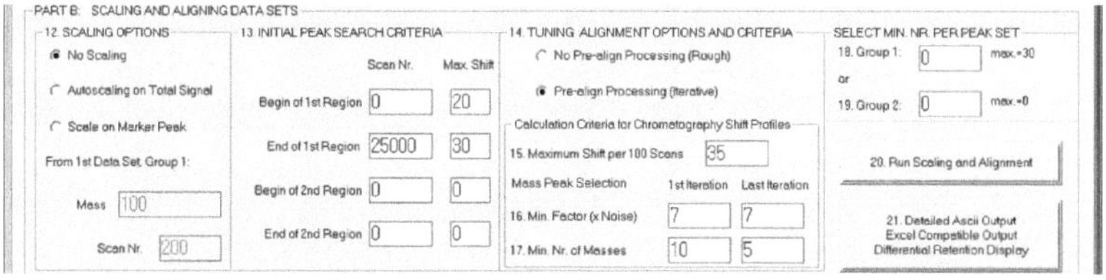

Figure 4.14 Screenshot of Part B: data processing parameters for peak alignment in metAlign. (Reprinted with permission from Dr. Arjen Lommen.)

experimental conditions described above are shown. However, as indicated, the max shift should be changed based on the data sets in the series. Check the chromatogram shifts as described above by using the RedMSViewer.

l. Click "20. Run Scaling and Alignment" button.

m. Click "21. Detailed Ascii Output, Excel Compatible Output, Differential Retention Display".

① Check the "Multivariate Compatible Output" button in the OUTPUT OPTIONS.

② Click "Make/View" button.

③ The End_result_amplitudes.csv is generated in the "2-1_abs" folder in the result folder.

④ The result csv file is utilized in AIoutput software.

(3) AIoutput data processing (Figure 4.15)

a. Download AIoutput from the RIKEN PRIMe web site. (Select either 32- or 64-bit version.) Do not change any Excel file names in the download folder.

b. Open "AIoutput.xlsm" file. Click the "AIoutput2" button in RawTable sheet.

Note: Depending on the PC condition, the user may get an error "Compile Error: Can't find project or library". In this case, go to "Tools" = > "References" and confirm the activation of "Microsoft Excel 12.0 Object Library" and "Microsoft DAO 3.5 or 3.6 Object Library".

c. File import

① If previous data already exist, clear the data by clicking "Data Clear" button.

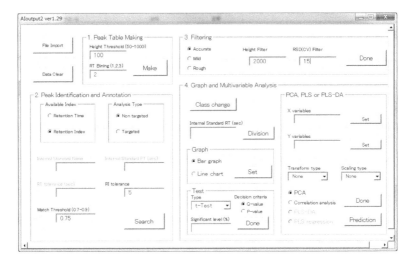

Figure 4.15 Screenshot of AIoutput. (Reprinted with permission from Dr. Hiroshi Tsugawa.)

② Select the result file "End_result_amplitudes.csv" derived from MetAlign "2-1_abs" folder.

d. 1. Making a Peak Table

The details of each parameter are written in the AIoutput manual. Here, the suitable parameters for the above experimental conditions are shown.

① Add 100 and 2 for "Height Threshold" and "RT Binning", respectively.
② These parameters are utilized for spectra extraction, that is, component detection. AIoutput treats the retention time as a second timescale. The program regards the mass trace information of the same retention time range combined with the binning parameter as a single component (metabolite). The binning parameter means the number of decimal places to merge the retention time. The height parameter is a threshold value. If the highest intensity of a single component (i.e., mass spectra) is less than this threshold, the component is not extracted.
③ Click "Make" button.

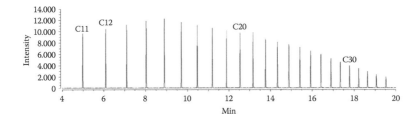

Figure 4.16 Chromatogram of the alkane mixture. (Reprinted with permission from Dr. Hiroshi Tsugawa.)

④ The result is generated in "RawTable", "MZTable", and "PeakTable" sheets.

e. 2. Peak Identification and Annotation

Here, the compound identification with retention index information is introduced.

① Check the analysis data of alkane mix and its retention time. Note that the molecular m/z of *n*-alkane (C10: m/z 142, C11: m/z 156, C12: m/z 170, etc.) can be seen (Figure 4.16).
② Fill in the retention time information of each alkane to "Alkane_Mix.xlsx" file. Keep the format (Figure 4.17).
③ Check the "Retention Index" in "Available Index" and also check the "Nontargeted" in "Analysis Type".
④ It is recommended to set the RI tolerance in the range of 5 to 10 and the Match Threshold in the range of 0.75 to 0.9.
⑤ Click "Search" button.
⑥ The result is generated in "PeakTable" sheet (Figure 4.18).

f. Classification of samples (Figure 4.19)
① Divide all samples into the same categories depending on your experiment.
② The class index must be added in the "PeakTable" sheet as shown below.
③ The order of class number is flexible. However, numbers must be used from 1 without skipping.

	A	B
1	Name	RT
2	C10_Alkane	252.093
3	C11_Alkane	319.693
4	C12_Alkane	384.593
5	C13_Alkane	444.793
6	C14_Alkane	500.493
7	C15_Alkane	552.993
8	C16_Alkane	601.343
9	C17_Alkane	647.443
10	C18_Alkane	691.143
11	C19_Alkane	732.793
12	C20_Alkane	772.143
13	C21_Alkane	809.793
14	C22_Alkane	845.843
15	C23_Alkane	880.393
16	C24_Alkane	913.593
17	C25_Alkane	945.443
18	C26_Alkane	976.093
19	C27_Alkane	1005.643
20	C28_Alkane	1034.195
21	C29_Alkane	1061.695
22	C30_Alkane	1088.295
23	C31_Alkane	1114.195
24	C32_Alkane	1139.295
25	C33_Alkane	1165.295
26	C34_Alkane	1193.745
27	C35_Alkane	1225.645
28	C36_Alkane	1261.745
29	C37_Alkane	1303.545
30	C38_Alkane	1350.845
31		
32		

Figure 4.17 Contents of Alkane_Mix.xlsx. (Reprinted with permission from Dr. Hiroshi Tsugawa.)

④ AIoutput utilizes the class index to filter "Unknown" metabolites and add the same color for statistical analysis such as *t* test, principal component analysis, and so on.

g. Filtering of unknown metabolites (Figure 4.20)
① Click "Done" button.
② The result is generated in the "PeakTableUpDate" sheet as shown below.

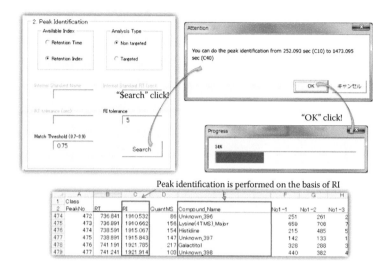

Figure 4.18 How to do the peak identification in AIoutput program. (Reprinted with permission from Dr. Hiroshi Tsugawa.)

Figure 4.19 How to add the class index in AIoutput program. (Reprinted with permission from Dr. Hiroshi Tsugawa.)

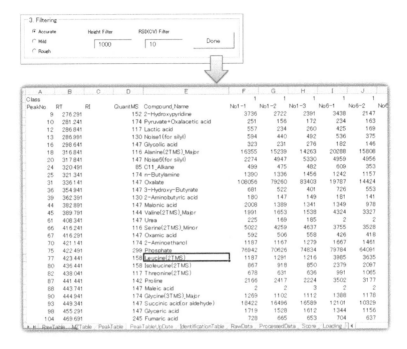

Figure 4.20 How to do the filtering method and how to get an organized data matrix in AIoutput program. (Reprinted with permission from
Dr. Hiroshi Tsugawa.)

Note: AIoutput uses the filtering method to extract only
meaningful information from unknown metabolites.

AIoutput can perform three filtering methods. Here, only
the suitable parameters are shown below. Accurate mode,
Height Filter: 5000, and RSD Filter: 10. The detail is shown
in the AIoutput manual.

Note that the filtering method is performed only for
unknown metabolite information.

h. Statistical analysis

AIoutput implements some statistical analysis methods
such as graph analysis, multi-t-test, principal component
analysis, correlation analysis, projection to latent structure
regression and projection to latent structure discriminant

analysis. The details of statistical analysis in AIoutput are described in Chapter 6.

4.5 Summary

GC/MS-based metabolome analysis is high throughput and highly reproducible. The total time required from sample derivatization to an organized data matrix construction for 100 samples is about 5 days (derivatization and density adjustment of samples: 1 day, instrument analysis: 3 days, data processing: 3 hours). Using the described methodology, the analysis time can be shortened making the whole pipeline highly efficient. Thus, the researcher can spend more time and effort in contemplating the biological understanding of the study.

In addition, two of the most important challenges in GC/MS-based metabolomics are to increase the number of identified metabolites and develop reliable criteria for identification of results. Just recently, these problems have been properly addressed by the groups of Fiehn,[19] Hummel,[20] and Matsuda.[21] The challenges are worth continuing.

Our task is to familiarize researchers in other fields of study with metabolomics-based study. In order to do this, we are developing more useful, easier, and reliable platforms.

References

1. Hiller, K., Hangebrauk, J., Ja, C., Spura, J., Schreiber, K., and Schomburg, D., *Analytical Chemistry*, 2009, 81: 3429–3439.
2. Tsugawa, H., Bamba, T., Shinohara, M., Nishiumi, S., Yoshida, M., and Fukusaki, E., *Journal of Bioscience and Bioengineering*, 2011, 112: 292–298.
3. Fiehn, O., Kopka, J., Dörmann, P., Altmann, T., Trethewey, R. N., and Willmitzer, L., *Nature Biotechnology*, 2000, 18(11): 1157–1161.
4. Pongsuwan, W., Bamba, T., Harada, K., Yonetani, T., Kobayashi, A., and Fukusaki, E., *Journal of Agricultural and Food Chemistry*, 2008, 56: 10705–10708.

5. Sreekumar, A., Poisson, L. M., Rajendiran, T. M., Khan, A. P., Cao, Q., Yu, J., Laxman, B., Mehra, R., Lonigro, R. J., Li, Y., Nyati, M. K., Ahsan, A., Kalyana-Sundaram, S., Han, B., Cao, X., Byun, J., Omenn, G. S., Ghosh, D., Pennathur, S., Alexander, D. C., Berger, A., Shuster, J. R., Wei, J. T., Varambally, S., Beecher, C., and Chinnaiyan, A. M., *Nature,* 2009, 457: 910–914.

6. Nishiumi, S., Kobayashi, T., Ikeda, A., Yoshie, T., Kibi, M., Izumi, Y., Okuno, T., Hayashi, N., Kawano, S., Takenawa, T., Azuma, T., and Yoshida, M., *PLoS ONE,* 2012, 7: e40459.

7. Kind, T., Wohlgemuth, G., Lee, D. Y., Lu, Y., Palazoglu, M., Shahbaz, S., and Fiehn, O., *Analytical Chemistry,* 2009, 81: 10038–10048.

8. Horai, H., Arita, M., Kanaya, S., Nihei, Y., Ikeda, T., Suwa, K., Ojima, Y., Tanaka, K., Tanaka, S., Aoshima, K., Oda, Y., Kakazu, Y., Kusano, M., Tohge, T., Matsuda, F., Sawada, Y., Hirai, M. Y., Nakanishi, H., Ikeda, K., Akimoto, N., Maoka, T., Takahashi, H., Ara, T., Sakurai, N., Suzuki, H., Shibata, D., Neumann, S., Iida, T., Tanaka, K., Funatsu, K., Matsuura, F., Soga, T., Taguchi, R., Saito, K., and Nishioka, T., *Journal of Mass Spectrometry,* 2010, 45: 703–714.

9. Schauer, N., Steinhauser, D., Strelkov, S., Schomburg, D., Allison, G., Moritz, T., Lundgren, K., Roessner-Tunali, U., Forbes, M. G., Willmitzer, L., Fernie, A. R., and Kopka, J., *FEBS Letters,* 2005, 579: 1332–1337.

10. Roger, A. L. and Charles, C. S., *Analytical Biochemistry*, 1971, 43: 533–538.

11. Kenneth R. L., Roy, H. R., and Charles, W. G., *Journal of Chromatography*, 1977, 141: 355–375.

12. Kanani, H. H. and Klapa, M. I., *Metabolic Engineering,* 2007, 9: 39–51.

13. Tsugawa, H., Tsujimoto, Y., Arita, M., Bamba, T., and Fukusaki, E., *BMC Bioinformatics,* 2011, 12: 131.

14. Lommen, A. and Kools, H. J., *Metabolomics,* 2012, 8: 719–726.

15. Lommen, A. 2009, *Analytical Chemistry*, 81: 3079–3086.

16. Fiehn, O., Kopka, J., Trethewey, R. N., and Willmitzer, L., *Analytical Chemistry,* 2000, 72: 3573–3580.

17. Fukusaki, E., Jumtee, K., Bamba, T., and Yamaji, T. *Zeitschrift für Naturforschung,* 2006, 267–272.

18. Dunn, W. B., Broadhurst, D., Begley, P., Zelena, E., Francis-McIntyre, S., Anderson, N., Brown, M., Knowles, J. D., Halsall, A., Haselden, J. N., Nicholls, A. W., Wilson, I. D., Kell, D. B., and Goodacre, R., *Nature Protocols,* 2011, 6: 1060–1083.

19. Abate, S., Ahn, Y. G., Kind, T., Cataldi, T. R. I., and Fiehn, O., *Rapid Communications in Mass Spectrometry,* 2010, 7: 1172–1180.
20. Hummel, J., Strehmel, N., Selbig, J., Walther, D., and Kopka, J., *Metabolomics,* 2010, 6: 322–333.
21. Matsuda, F., Tsugawa, H., and Fukusaki, E., *Analytical Chemistry,* 2013, 85: 8291–8297.

5

LC/QqQ/MS Analysis

*Widely Targeted Metabolomics
on the Basis of Multiple
Reaction Monitoring*

Chapter 5
LC/QqQ/MS Analysis

Fumio Matsuda and Hiroshi Tsugawa

Chapter Outline

5.1 Introduction

The general aim of metabolomics is to analyze comprehensively all known and unknown metabolites in biological samples.[1] The first generation of metabolomics research mainly focused on the development of a methodology for nontargeted metabolome analysis to analyze all detectable metabolites without the preselection of targets. To detect a wider range of metabolites, nontargeted metabolome analyses have been performed while operating the mass spectrometer in full scan mode. Time-of-flight (TOF) analyzers are suitable for this purpose because of their higher scan speed and sensitivity.[2–8] The high resolution of modern TOF analyzers is also preferred in order to characterize unknown metabolites by using accurate mass data.[9,10] The nontargeted metabolome analysis strategy, using LC-TOF-MS, plays a critical role in the metabolomics research of plant secondary metabolites, because almost all detectable metabolite signals are unknown phytochemicals.[11] However, in many biological applications, multitargeted or widely targeted data analyses have been performed using the raw data produced by nontargeted methodologies. The return to targeted analysis is due to the difficulty in the structural characterization of unknown metabolite signals based only on high-resolution and tandem mass spectral data.[12,13] For analyses exploring the behavior of metabolic systems, the study can be performed using a data set of identified metabolites without considering the unknown metabolites.[14]

For widely targeted metabolome analyses, there is another approach to consider, which employs triple-stage quadrupole mass spectrometers (QqQ/MS). The QqQ/MS was developed for the sensitive analysis of several preselected metabolites by operating in selected or multiple reaction monitoring (S/MRM) mode. In the S/MRM mode, the first (Q1) and third (Q3) quadrupoles serve as mass filters. Precursor ions selected in Q1 are dissociated in the collision cell of Q2 and the generated fragments are passed through to Q3. By selecting suitable pairs of m/z values for Q1 and Q3 and optimizing the energy for the collision-induced dissociation, selective and sensitive detection of a target metabolite is achieved. Recent advances in terms of improved sensitivity (zepto mole) and higher operation speed (less than one millisecond (ms) for one channel) of QqQ/MS instrumentation enables the simultaneous analyses of widely targeted metabolites. Because 0.2 s/point intervals are required for the accurate determination of the peak area in chromatograms, up to 100 metabolites (1 ms * 100 metabolites * (1 quantification and 1 confirmation ion) = 200 ms) can be analyzed in one time segment. Current QqQ/MS systems equipped with the functionality to control the time segments dynamically, such as scheduled MRM and programmed SRM, increase the number of metabolites analyzed in one run to several hundreds. In this chapter, the current status and protocols for widely targeted metabolomics on the basis of multiple reaction monitoring are introduced.

5.2 Advantages and Disadvantages of LC/MS Analysis

In widely targeted metabolome analyses, reproducibility of the retention time for each metabolite and sharpness of peak shapes are especially important for accurate execution of a scheduled MRM program. There are several separation methods that can be used with QqQ/MS for widely targeted metabolomics analyses. In this regard, the use of gas chromatography–triple-stage quadrupole mass spectrometers (GC/QqQ/MS) is a promising method because of its excellent peak capacity and ability to convert retention time to retention index (RI) by using n-alkane data.[15] Although the use of capillary electrophoresis triple-stage quadrupole mass spectrometers (CE/QqQ/MS) is a very powerful method for widely targeted analyses of anionic and cationic metabolites, the instability in migration time across analytical runs may negate the advantage of the QqQ/MS.[16] In this chapter, we focus on widely targeted metabolomics using liquid chromatography (LC) as the separation method. LC/QqQ/MS can be an interesting platform that takes advantage of the compatibility in QqQ/MS and flexibility in the LC methods. Moreover, technological developments in LC separation of sugar phosphates and cofactors have expanded the possibility of widely targeted metabolomics analyses using LC/QqQ/MS.

A set of information is required for the specific analysis of one metabolite, that is, retention time and an MRM method for selective detection by QqQ/MS. For metabolites such as L-alanine, which is identical in both microorganisms and mammals, the detection methods developed for each universal metabolite can be applied for all species and biological samples without any modification. Among the variable parameters used in the analytical method, the MRM method is specific to a QqQ/MS model and does not essentially depend on the LC method. It has been considered that MRM methods can be adapted among QqQ/MS models of the same manufacturer, suggesting that developed MRM methods for each metabolite can be shared across different laboratories. As reported in several publications, MRM data sets can facilitate widely targeted metabolome analyses without the need to develop specific, individual MRM methods. For instance, MRM conditions for 555 metabolites developed using a Waters MS/MS system have been published.[17] This compatibility in detection methods is the first advantage of widely targeted metabolomics using QqQ/MS. Furthermore, in widely targeted metabolomics, target metabolites are not only known metabolites with available standards but also known–unknown metabolites. It has been reported that there are many signals of unknown metabolites that are always observed in the metabolome data obtained from samples of the same tissue or species. Although the structure of these metabolites remains unknown, the metabolites can be analyzed by QqQ/MS using a MRM method developed based on the MS/MS spectral data of the metabolite signals.[18] Thus, widely targeted metabolomics can theoretically deal with an unlimited number of metabolites by using a collection of MRM methods for many known–known and known–unknown metabolites.

The retention time of a metabolite depends on the LC method. Although there are multiple parameters that determine the LC method, such as column length (from 1 cm to 100 cm), bore (0.5–4.6 mm), chemistry of the stationary phase, particle size (sub 2 micron–5 micron), mobile phase, and its gradient curve, the most important parameter in metabolomics is the runtime required for one injection. For analyzing a large number of metabolites in a sample extract, there are two strategies for method development. The first strategy is to develop an LC method with a relatively long gradient elution program (15–60 min) using long columns (75, 150, and 250 mm). This strategy is employed in the analysis of sugar phosphates in order to separate several structural isomers, such as glucose-6-phosphate and fructose-6-phosphate, whose MRM conditions are identical. The distinct analysis of glucose-6-phosphate and fructose-6-phosphate is critical in certain studies, such as those concerned with the understanding of metabolic systems.[19-21] However, such distinct separation is not always required, such as in high-throughput phenotype screens of a large number of bioresources.[22] The second strategy is to employ a very short analytical runtime (2–5 min/injection) using a ballistic gradient curve.[17] Although this strategy ignores the separation of some isomers, the technological advantage is the potential for high-throughput analysis and a customizable experimental design. For instance, analyses of 800 samples can be finished within two days using a customized method for 100 core metabolites with a 3.0-min runtime. Based on the metabolome data, 100 positive samples can then be selected. An additional 10 injections are performed for each sample by employing 10 methods for the analysis of the 1,000 metabolites (100 metabolites analyzed 10 times), which can finish in three days. The flexibility in method development is an added advantage of widely targeted metabolomics using LC/QqQ/MS.

Intermediates and cofactors related to the central carbon metabolism, such as glucose-6-phosphate, pyruvate, succinate, ATP, and NADH are the most interesting targets in biological analyses because of their important roles in metabolism. However, comprehensive analysis of these intermediates has been difficult because of poor chromatographic separation caused by their hydrophilic and anionic characteristics. The development of a CE/MS method was a big breakthrough for nontargeted profiling of these intermediates. The method invented by Luo et al.[20] employing the commonly used LC/QqQ/MS setup is the next important step for a widely targeted metabolome analysis. The separation of several isomers, such as glucose-6-phosphate and fructose-6-phosphate is achieved by the introduction of ion-pair LC using tributylamine as the ion-pairing reagent. The method has been improved for more stable separation of sugar phosphates and applied for nontargeted metabolome analyses of microorganisms and mammals.

A protocol for the method is described below including some technical tips.

5.3 Analytical Method for LC/ QqQ/MS Analysis

Luo et al. (2007) originally developed the method introduced here. The MRM conditions were modified for Shimadzu and Agilent QqQ/MS models. The method has been used in many studies and several modifications have been reported.

5.3.1 Reagents

- Buffer A: 15 mM tributyl amine (TBA), 10 mM acetic acid in water (LC/MS grade)
- Buffer B: Methanol (LC/MS grade)
- Tributyl amine (TBA): SIGMA-ALDRICH, No. 471313-250 mL
- Acetic acid: Wako, LC/MS grade, No. 010-19112
- Methanol: Wako, LC/MS grade, No. 210-01303
- Ultra pure water: Wako, LC/MS grade, No. 134-14523

5.3.2 Preparation of Buffer A

1. Measure 500 mL of ultra pure water using a clean graduated cylinder.
2. Transfer approximately 50 mL of ultra pure water to a solvent bottle for buffer A.
3. Add 450 µL acetic acid and then 1190 µL of TBA to the ultra pure water in solvent bottle A.
4. Mix well by sonication until the solution is homogeneous (oily TBA disappears).
5. Add the remaining ultra pure water (450 mL) to the solution and mix well.

Tip: The composition of buffer A drastically affects the retention times of sugar phosphates. Prepare fresh buffer A at least every two days.

5.3.3 *Apparatus*

- Column: ProteCol™ C18 Q103, 150 × 2.1 mm, 100 Å, 3 μm (SGE Analytical Science, No. 2C183-03N02P), L-column 2 ODS, 2.1 × 150 mm, 3 μm (CERI No. 711020)
- Glass vial 12 × 32 mm: (CHROMACOL LTD, No. 500*02-SV)
- Glass insert vial: (CHROMACOL LTD, No. 1000*02-MTV)
- Cap: (CHROMACOL LTD No. 500*8-SCJ(W)00139321)
- Septum: (CHROMACOL LTD No. 500*8-ST101 96653)

5.3.4 *LC/MS Conditions for LCMS-8040 (Shimadzu)*

5.3.4.1 *HPLC Conditions*

- Column: 3L-column2 ODS 2.1 × 150 mm, 3 μm (CERI No. 711020)
- Flow rate: 0.3 ml/min
- Column temperature: 40°C
- Injection volume: 3 μL
- Gradient curve

min	A (%)	B (%)
0.5	100	0
7.5	75	25
11	10	90
12	10	90
12.1	100	0
15	100	0

5.3.4.2 *MS Conditions*

- Probe and detection mode: ESI, negative, MRM mode
- Probe position: +1.5 mm
- DL temperature: 250°C
- Nebulizer gas flow: 2 L/min
- Heat block temperature: 400°C
- Conditions for other methods were determined by using the Auto-tuning function.

5.3.4.3 MRM Conditions (Table 5.1)

5.3.5 LC/MS Conditions for Agilent 6460

5.3.5.1 HPLC Conditions

- Column: ProteCol™ C18 Q103, 150 × 2.1 mm, 100Å, 3 µm (SGE Analytical Science, No.2C183-03N02P)
- Flow rate: 0.3 ml/min
- Column temperature: 35°C
- Injection volume: 3 µL
- Gradient curve

min	A (%)	B (%)
0	100	0
5	100	0
24	10	90
24.1	100	0
30	100	0

5.3.5.2 MS Conditions

- Probe and detection mode: ESI, negative, MRM mode
- Gas temp: 300 C°
- Gas flow: 10 L/min
- Nebulizer: 55 psi
- Sheath gas temp: 380 C°
- Sheath gas flow: 11 L/min
- Capillary: 3,500 V
- Nozzle voltage: 1,000 V
- Resolution: Q1/Q3, wide/wider
- EM voltage: 2,000 V

5.3.5.3 MRM Conditions (Table 5.2)

TABLE 5.1
MRM Conditions (1) in Ion-Pair LC/MS/MS

No.	Name	Retention Time (min)	MRM	Target Q1 Pre Bias	Target Collision Energy	Target Q3 Pre Bias
1	Glycine	0.001	74.00>74.00	10	20	14
2	Arginine	0.905	173.10>131.20	13	15	24
3	Histidine	0.907	154.00>93.15	12	21	16
4	4-Aminobutyrate	1.037	102.00>84.00	17	15	15
5	Serine	1.138	104.00>74.15	12	16	13
6	Asparagine	1.151	131.00>113.15	10	15	21
7	Glutamine	1.16	145.00>127.05	12	18	18
8	Hydroxyproline	1.179	130.00>84.15	10	14	15
9	Homoserine	1.182	118.00>100.00	11	17	18
10	Ornithine	1.183	131.10>85.00	12	15	17
11	Threonine	1.186	118.00>74.05	11	15	13
12	Leucine	1.201	130.10>84.00	11	15	15
13	Ribitol	1.243	151.00>89.10	11	15	16
14	Cysteine	1.246	120.00>33.00	11	10	29
15	2-Aminobutyrate	1.258	102.00>45.00	17	14	16
16	Trehalose	1.302	341.00>89.00	15	23	16
17	Proline	1.331	114.00>68.10	10	15	11
18	D-Glucono-1,5-lactone	1.533	177.00>129.00	13	15	29
19	Valine	1.554	116.10>45.00	10	16	16
20	Cytidine	1.845	242.00>109.15	19	15	19
21	Methionine	1.987	148.00>47.05	11	14	16
22	Theanine	2.006	173.00>155.25	13	17	29
23	Guanine	2.499	150.00>133.10	11	20	24
24	Isoleucine	2.578	130.10>45.00	11	15	15
25	Tyrosine	2.834	180.00>163.05	12	18	18
26	Amino adipic acid	3.345	160.00>116.20	12	17	21
27	Glutamate	3.477	146.00>102.20	11	15	18
28	Uridine	3.548	243.00>110.15	19	18	20
29	Aspartate	3.685	132.00>88.05	10	14	15

(continued)

TABLE 5.1 (CONTINUED)
MRM Conditions (1) in Ion-Pair LC/MS/MS

No.	Name	Retention Time (min)	MRM	Target Q1 Pre Bias	Target Collision Energy	Target Q3 Pre Bias
30	Thymine	4.003	125.00>42.05	10	14	14
31	Inosine	4.559	267.00>135.15	21	23	24
32	Guanosine	4.706	282.10>150.20	22	21	28
33	Phenylalanine	4.854	164.00>103.15	13	18	19
34	Shikimate	4.905	173.00>93.15	13	19	17
35	Glycerate	5.088	105.00>75.15	12	15	26
36	Thymidine	5.249	241.10>42.05	18	17	14
37	Glycolate	5.336	75.00>75.00	16	15	15
38	Glyoxylate	5.577	73.00>73.00	14	13	15
39	Ascorbate	5.639	175.00>115.00	11	14	21
40	Myo-inositol	5.861	179.00>96.90	14	12	17
41	Glucosamine	6.064	178.00>142.00	18	15	26
42	Inositol	6.34	179.00>87.00	14	20	15
43	Lactate	6.348	89.00>43.10	10	15	14
44	Hexose	6.356	179.00>89.05	14	9	16
45	Pyroglutamate	6.438	128.00>84.10	10	14	15
46	G6P	6.444	258.90>97.05	20	21	17
47	PIPES	6.637	301.00>193.25	12	28	21
48	R5P	6.759	229.10>96.95	18	13	18
49	S7P	6.867	288.90>97.10	23	23	17
50	F6P	6.927	258.90>97.10	20	15	17
51	Tryptophan	6.978	203.10>116.15	16	18	21
52	α-Glycerophosphate	7.074	171.10>79.10	13	16	13
53	G1P	7.113	258.90>79.05	21	28	27
54	Glutathione	7.138	305.90>143.20	21	19	26
55	GAP	7.313	168.90>97.10	12	11	17
56	E4P	7.5	198.90>97.20	13	12	17
57	Ru5P	7.622	229.00>97.10	17	13	17
58	β-Glycerophosphate	7.753	170.90>79.05	13	19	13
59	Orotate	7.84	155.00>111.15	12	14	20

TABLE 5.1 (CONTINUED)
MRM Conditions (1) in Ion-Pair LC/MS/MS

No.	Name	Retention Time (min)	MRM	Target Q1 Pre Bias	Target Collision Energy	Target Q3 Pre Bias
60	F1P	7.925	258.90>97.05	20	21	17
61	CMP	7.964	322.00>79.10	25	28	13
62	NAD	8.243	662.10>540.10	26	18	26
63	Alanine	8.268	88.00>44.00	14	14	15
64	Pyruvate	8.275	87.00>43.05	10	11	14
65	DHAP	8.391	168.90>97.05	13	12	17
66	UMP	8.661	322.90>97.10	25	24	17
67	GMP	8.995	362.00>79.10	29	28	13
68	Oxalacetate	9.31	131.00>87.00	25	11	27
69	TMP	9.719	321.00>79.10	25	38	14
70	AMP	9.811	346.00>79.05	14	32	14
71	Nicotinate	9.983	122.00>78.15	14	15	13
72	Pantothenate	10.022	218.00>88.00	21	14	16
73	Succinate	10.155	117.00>73.20	13	15	12
74	Fumarate	10.278	115.00>71.10	13	10	12
75	cAMP	10.465	328.00>134.10	15	27	24
76	Malate	10.578	132.90>115.20	10	17	21
77	Lysine	10.653	145.00>87.00	14	14	13
78	UDP-Glu	10.712	564.80>323.10	24	26	15
79	2-Oxoglutarate	10.745	145.00>101.20	11	13	18
80	CDP	10.753	401.80>79.05	16	43	13
81	6-Phosphogluconate	10.771	275.00>79.00	11	8	24
82	GDP	10.806	442.00>79.10	18	45	13
83	ADP-Glu	10.806	588.00>346.15	24	26	24
84	UDP	10.807	402.90>79.05	16	48	13
85	ADP-Rib	10.81	558.00>346.15	22	26	16
86	NADP	10.811	741.80>620.10	26	18	30
87	KDPG	10.828	256.90>97.05	10	18	17

(continued)

TABLE 5.1 (CONTINUED)
MRM Conditions (1) in Ion-Pair LC/MS/MS

No.	Name	Retention Time (min)	MRM	Target Q1 Pre Bias	Target Collision Energy	Target Q3 Pre Bias
88	3PGA	10.829	184.90>97.05	14	16	17
89	F2,6P	10.834	338.90>241.15	26	19	27
90	F1,6P	10.838	338.90>97.10	26	22	17
91	NADH	10.876	664.00>78.95	24	57	13
92	RuBP	10.887	308.90>97.05	24	20	17
93	Isocitrate	10.891	190.90>73.20	13	22	26
94	Citrate	10.892	190.90>87.00	13	18	14
95	ADP	10.913	425.90>79.10	17	47	13
96	1,3-BPG	10.919	265.00>167.15	11	18	29
97	Phosphoenolpyruvate	10.928	167.00>78.95	15	13	13
98	FMN	10.985	455.00>97.10	18	30	17
99	2-Isopropylmalate	10.998	175.00>115.20	13	16	21
100	FAD	11.155	783.90>97.10	20	51	17
101	CTP	11.171	481.90>159.10	19	36	29
102	GTP	11.185	521.90>159.05	20	32	29
103	NADPH	11.201	744.00>159.00	26	60	30
104	UTP	11.206	482.90>159.10	19	36	29
105	ATP	11.226	505.90>159.10	20	35	29
106	CoA	11.343	766.50>79.00	20	54	13
107	Malonyl CoA	11.367	852.10>408.10	22	40	29
108	D-Camphorsulfonic acid	11.372	231.00>80.00	11	32	29
109	Acetyl CoA	11.382	808.00>408.00	20	37	28
110	Succinyl CoA	11.391	866.00>408.00	20	41	27

Source: Dr. Fumio Matsuda, unpublished data.

TABLE 5.2
MRM Conditions (2) in Ion-Pair LC/MS/MS

KEGGID	Name	Q1	Q3	Fragmenter	Collision Energy	Retention Time
	D-camphor-10-sulfonic acid	231	80	130	33	19.4
C00048	Glyoxylate	73	45	50	1	6.8
C00160	Glycolate	75	47	50	6	7.2
C00022	PYR	87	43	50	3	13.3
C00186	Lactate	89	43	50	7	10
C00209	Oxalate	89	43	65	5	10
C00164	AcAc	101	57	50	5	12.6
C00109	2-Oxobutyrate	101	57	54	1	15.8
C01089	3-Hydroxybutyrate	103	59	50	2	11.4
C00258	Glycerate	105	75	50	8	7.3
C00380	Cytosine	110	67	35.1	8	1.2
C00106	Uracil	111	42	50	10	2
C00122	Fumarate	115	71	50	1	17.4
C06255	2-Oxovalerate	115	71	54	1	17.6
C00042	Succinate	117	73	50	5	16.1
C00253	Nicotinate	122	78	50	7	15.6

(continued)

TABLE 5.2 (CONTINUED)
MRM Conditions (2) in Ion-Pair LC/MS/MS

KEGGID	Name	Q1	Q3	Fragmenter	Collision Energy	Retention Time
C00178	Thymine	125	42	50	10	4
C00300	Creatine	130	88	54	1	1.3
C00152	Asn	131	96	54	1	1.1
C00036	OXA	131	87	50	5	17.6
C00049	Asp	132	88	70.2	8	5.3
C04039	2,3-Dihydroxyisovalerate	133	57	54	13	14.5
C04039	2,3-Dihydroxyisovalerate	133	75	54	9	14.5
C00149	Malate	133	115	50	5	16.7
C00147	Adenine	134	107	90	13	4.7
C00262	Hypoxanthine	135	92	100	12	3
C00108	Anthranilic acid	136	92	54	13	17.4
C00026	AKG	145	101	50	5	17.1
C00064	Gln	145	127	63.7	6	1.2
C00025	Glu	146	128	78	6	–
C00242	Guanine	150	133	98.8	8	3
C00079	Phe	164	147	104	8	5
C00074	PEP	167	79	50	5	17.8

C00118	G3P	169	97	50	5	11
C00111	DHAP	169	97	50	5	13.1
C00623	Glycerol 1-phosphate	171	79	113.1	35	11.3
C00093	Glycerol3P	171	79	90	9	12.1
C00493	Shikimate	173	93	100	9	7.2
C00417	Aconitate	173	85	50	9	17.9
C00062	Arg	173	131	126.1	11	0.9
C00327	Citrulline	174	131	50	4	1.2
C02504	2-Isopropylmalic acid	175	115	60	9	17.9
C00082	Tyr	180	163	114.4	9	3
C00631	2PG	185	79	50	29	17.5
C00197	3PG	185	79	50	29	17.5
C00311	Isocitrate	191	73	50	17	17.6
C00158	Citrate	191	111	50	5	17.8
C00333	D-Galacturonate	193	59	60	12	
C00191	Glucuronate	193	59	50	12	5.8
C01494	Ferulic acid	193	134	72.8	13	17.7
C00257	D-Gluconate	195	75	80	20	6.9
C00279	E4P	199	79	50	29	12.7
C00279	E4P_2	199	97	50	5	12.7

(continued)

TABLE 5.2 (CONTINUED)
MRM Conditions (2) in Ion-Pair LC/MS/MS

KEGGID	Name	Q1	Q3	Fragmenter	Collision Energy	Retention Time
C00078	Trp	203	116	54	9	12
C00864	Pantothenic acid	218	88	90	5	15.7
C00542	L-Cystathionine	221	134	54	5	1.1
C00386	Carnosine	225	154	54	9	0.9
C00117	R5P	229	97	50	5	11.2
C00231	X5P	229	97	50	5	11.9
C00199	Ribu5P	229	97	50	5	12.1
C00491	L-Cystine	239	120	54	1	1.1
C00214	Thymidine	241	125	90	1	9.2
C00475	Cytidine	242	109	50	3	3.6
C00299	Uridine	243	110	90	5	10.5
C00092	G6P	259	79	90	40	11.3
C00085	F6P	259	97	90	9	11.3
C00085	F6P_2	259	169	90	2	11.5
C00103	G1P	259	241	90	2	12.8
C01094	D-Fructose 1-phosphate	259	79	65	40	11.1
C03384	D-Galactose 1-phosphate	259	241	65	8	

C00636	D-Mannose 1-phosphate	259	79	65	40	11.2
C00275	D-Mannose 6-phosphate	259	79	65	40	10.4
C00212	Adenosine	266	134	50	9	
	DL-Homocystine	267	132	54	1	1.5
C00345	6PG	275	79	90	29	17.3
C00387	Guanosine	282	150	150	9	
C05382	S7P	289	97	90	9	11.3
C00051	GSH	306	143	90	9	12.8
C01182	RuBP	309	97	50	17	17.6
C00364	dTMP	321	79	90	33	15.7
C05822	CMP	322	79	100	40	13.5
C00105	UMP	323	79	90	37	14.7
C00575	cAMP	328	134	130	17	17
C00354	F16P	339	97	90	17	13.1
C00942	cGMP	344	150	130	17	16.1
C00020	AMP	346	79	130	21	15.8
C00130	IMP	347	79	100	40	15
C00144	GMP	362	79	90	21	15
C00447	SBP	369	97	90	9	12
C00255	Riboflavin	375	255	90	9	16.7
C00021	S-Adenosyl-L-homocysteine	383	134	103	17	11.2

(continued)

TABLE 5.2 (CONTINUED)
MRM Conditions (2) in Ion-Pair LC/MS/MS

KEGGID	Name	Q1	Q3	Fragmenter	Collision Energy	Retention Time
C00119	PRPP	389	291	100	6	19.1
C00363	dTDP	401	79	100	40	18
C00015	UDP	403	111	75.4	15	17.6
C00008	ADP	426	79	250	40	18
C00035	GDP	442	150	130	21	17.7
C00061	FMN	455	97	130	25	18.5
C00458	dCTP	466	159	100	25	19
C00459	dTTP	481	159	100	25	19.4
C00063	CTP	482	159	150	25	19
C00075	UTP	483	159	100	30	19.2
C00131	dATP	490	159	150	25	19.4
C00002	ATP	506	159	130	33	19.4
C00044	GTP	522	159	100	35	19.2
C00301	ADP-ribose	558	346	130	21	17.8
C00029	UDP-glucose	565	323	130	21	17.4
C00498	ADPG	588	346	130	21	17.8
C00394	GDPG	604	362	130	21	17.4
C00127	GSSG	611	306	130	21	16.9

C00003	NAD	662	540	90	13	14.4
C00004	NADH	664	79	170	40	18.1
C00006	NADP	742	620	90	9	17.6
C00005	NADPH	744	408	170	33	19.3
C00010	CoA	766	408	200	30	19.9
C00016	FAD	784	437	170	25	19.4
C00024	AcCoA	808	408	170	33	20.1
C00877	Crotonoyl CoA	834.6	408	201	29	20.9
C00136	Butyryl CoA	836.6	408	201	20	21.1
C00332	AcAcCoA	850	766	170	25	20
C01144	DL-beta-Hydroxybutyryl CoA	852.6	408	201	33	20.2
C00091	Succinyl CoA	866	426	200	35	20.1
C00227	Acetylphosphate	139	79	58.5	3	16.7
C00141	2-Ketovaline	115	71	65	3	17.2
C00490	Itaconate	129	85	46.8	5	17
C00489	Glutarate	131	113	57.2	5	16.2
C00805	Salicylate	137	93	63.7	12	19
C00418	Mevalonate	147	59	66.3	6	12.9
C05593	3-Hydroxyphenylacetate	151	107	67.6	4	17.6

(continued)

TABLE 5.2 (CONTINUED)
MRM Conditions (2) in Ion-Pair LC/MS/MS

KEGGID	Name	Q1	Q3	Fragmenter	Collision Energy	Retention Time
C01606	Phthalate	165	77	52	10	18.8
C00366	Urate	167	124	115.7	9	7.5
C00198	Gluconolactone	177	129	68.9	4	1.3
C00392	Mannitol	181	71	96.2	15	1.2
C00140	N-Acetyl-D-glucosamine	220	59	84.5	13	1.1
C00644	D-Mannitol 1-phosphate	261	79	117	34	10.8
C02170	Methylmalonate	117	73	58.5	5	16.5
C00180	Benzoic acid	121	77	63.7	7	19.2
C14418	3-Nitrophenol	138	108	58.5	9	20.3
C05629	3-phenylpropionic acid	149	105	91	7	20.3
C02518	Aminosalicylic acid	152	108	76.7	10	16
C00230	3,4-Dihydroxybenzoate	153	123	111.8	8	20.3
C00811	4-Coumarate	163	119	52	12	17.7
C01456	Tropate	165	103	28.6	6	17.8
C00738	Hexose	179	59	79.3	11	1
C02225	2-Methylcitrate	205	125	44.2	9	18.3
C00352	D-Glucosamine 6-phosphate	258	79	46.8	35	1.8
C01159	Glycerate diP	265	167	52	9	19.3

C00239	dCMP	306	79	161.2	46	14.2
C00365	dUMP	307	195	106.6	10	15
C00341	Geranyl diphosphate	313	79	88.4	36	20.5
C00360	dAMP	330	134	176.8	27	16.2
C00362	dGMP	346	133	156	38	15.3
C00705	dCDP	386	159	74.1	22	17.8
C00206	dADP	410	159	156	26	18
C00361	dGDP	426	275	137.8	17	17.9

Source: Kato et al., *Journal of Bioscience and Bioengineering,* 2012, 113: 665–673.

5.4 Representative Chromatograms

An extract prepared from *Saccharomyces cerivisiae* was analyzed. See Figure 5.1. Also please see Chapter 3 for sample preparation procedures, that is, cultivation, quenching, and extraction.

5.4.1 Troubleshooting

Tip 1: Because tributyl amine (TBA) is very difficult to wash out completely, the TBA signals (*m/z* 186 in positive ion mode)

Figure 5.1 Chromatograms by means of multiple reaction monitoring based LC/MS/MS. (From Dr. Fumio Matsuda, unpublished data.)

are observed in all analyses on the LC/MS system once the method is employed. In order not to contaminate other LC/MS systems with TBA, only specific bottles and columns are exclusively used for the method containing TBA. A water/acetonitrile/formic acid mixture in a 50:50:2 ratio is effective for washing out the TBA, as far as we tested.

Tip 2: The target metabolites of the method have various functional groups that interact with metals, which can cause difficulties in separation. Thus, the method often shows a compatibility problem with the column used. In this protocol, LC columns containing extremely pure silica (low metal) or stainless-free columns are employed because of their good performance. If serious tailings are observed for many peaks when using these columns, consider the interaction of the samples with the tubing of the LC system. Replacement of metal tubing with PEEK™ tubing and washing with aqueous phosphate sometimes show good improvement in the result.

5.5 Data Processing of LC/QqQ/ MS: MRMPROBS

The MRM-based data processing is dependent on manual analysis by means of the MS vendor software. The MS vendor's software is very useful, but sometimes it is difficult to construct an organized data matrix from the MRM raw data sets. Because data assessment usually relies on manual evaluation due to the lack of automated probabilistic measures, a large amount of time and effort is spent in constructing the data matrix. In addition, manual verification is always subjective, erroneous, and even irreproducible. Therefore, objective evaluation is needed to minimize misinterpretations of the results from the biological samples. Moreover, the developed platform must be able to process MRM data sets from any MS vendor.

To meet these requirements, we developed a software program written in C# called MRMPROBS, which is designed to process MRM data sets for widely targeted metabolomics.[23] The program evaluates the metabolite peaks by posterior probability, which is defined as the odds ratio by means of a newly optimized multivariate logistic regression model. The probability is calculated from the peak intensity, retention time and amplitude similarity, peak top differences, and peak shape similarity of a peak group compared to reference information. In addition, we also developed a user-friendly graphical interface to allow data organization and statistical analysis. MRMPROBS supports Shimadzu, Agilent, Thermo Fisher Scientific, and AB Sciex mass spectrometers. Thus, MRMPROBS is currently an optional platform for analysis of MRM data sets, although a more flexible and useful platform will be provided in the near future. Here, we give an introduction on how to use MRMPROBS for MRM-based metabolomics analysis.

Software

- MRMPROBS: http://prime.psc.riken.jp/(Follow the link: "Targeted and Non-targeted analysis software" = > "MRMPROBS")
- File converter: http://prime.psc.riken.jp/(Follow the link: "Targeted and Non-targeted analysis software" = > "MRMPROBS" = > File converter: Download)

Environment

- Windows OS (Recommended: Core i5 or more, RAM: 4 GB or more)
- .NET framework 3.5 or later
- LabSolutions installed for Shimadzu.lcd file
- MassHunter installed for Agilent.D file
- Analyst installed for AB Sciex.wiff and .wiff.scan file
- MSFileReader installed for Thermo Fisher Scientific. raw file

5.5.1 File Conversion (Figure 5.2)

1. Read license agreements and download the file converter.
2. Open "AnalysisBaseFileConverter.exe".
3. Drag and drop your analysis files (.lcd, .D, .wiff, or .raw files).
4. Click "Convert" button.

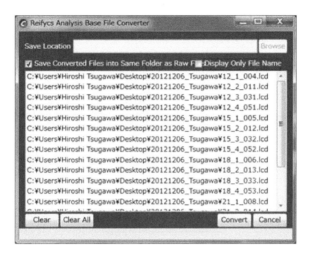

Figure 5.2 Screenshot of the file converter. (Reprinted with permission from Reifycs Inc.)

Note:

- Import just .wiff files for AB Sciex to the file converter. However, the .wiff.scan files need to be included in the same folder as the .wiff files.
- The converted .abf files are generated in the same folder as the raw data files.
- It should be noted that additional processes are sometimes required for the LabSolutions .lcd file conversion to .abf file. If error messages in the MRMPROBS program appear, download and read the manual for Shimadzu.lcd file conversion.

5.5.2 MRM Transitions Library (Figure 5.3)

1. Make a tab-separated format file including compound name, retention time, amplitude ratio, precursor m/z, and product m/z of targeted metabolites.

Note:

- The retention time for identical targeted metabolites should be the same.
- MRMPROBS recognizes "100" in the Ratio (%) column as the target transition used for quantification of

```
Supplemental data 1 (Example of MRM transitions library).txt - Notepad
File  Edit  Format  View  Help
Compound          RT(min) Ratio(%)          Precursor      Product
Arginine          1.46    100      173.1    131.2
Arginine          1.46    3.7      173.1    156.2
Arginine          1.46    1.8      173.1    41.1
Cytidine          2.54    100      242      109.15
Cytidine          2.54    14       242      42
Cytidine          2.54    11       242      152.2
Theanine          2.89    100      173      155.25
Theanine          2.89    34.2     173      84.2
Theanine          2.89    17.9     173      74.15
Guanine 3.42      100      150      133.1
Guanine 3.42      35.3     150      66.15
Guanine 3.42      26.5     150      108
2-Amino adipic acid        4.8     100      160      116.15
2-Amino adipic acid        4.8     105.9    160      142.15
2-Amino adipic acid        4.8     45.8     160      98
Uridine 4.95      100      243      110.15
Uridine 4.95      41.4     243      200.25
Uridine 4.95      29.1     243      42
Thymine 5.58      100      125      42.05
Inosine 6.57      100      267      135.15
Inosine 6.57      13.4     267      108.02
Inosine 6.57      6.6      267      92.03
Guanosine         6.73     100      282.1    150.2
Guanosine         6.73     17.7     282.1    133.15
Guanosine         6.73     10.6     282.1    108.02
Shikimate         6.88     100      173      93.15
Shikimate         6.88     39       173      73.15
Shikimate         6.88     34       173      111.2
Glycerate         7.15     100      105      75.15
Glycerate         7.15     56.8     105      59.1
Glycerate         7.15     67.1     105      45.05
Thymidine         7.53     100      241.1    42.05
Thymidine         7.53     11.9     241.1    151.1
Lactate 9.32      100      89       43.1
Lactate 9.32      47.1     89       45.05
Lactate 9.32      10.5     89       17.05
G6P      9.21     100      258.9    97.05
G6P      9.21     30.1     258.9    79.05
```

Figure 5.3 Example of an MRM transition library. (From Tsugawa et al., *Analytical Chemistry,* 2013, 85: 5191–5199.)

a targeted metabolite. This program can accept only one transition for metabolite quantification. The other transitions are recognized as qualifier transitions and used for confirmation of a target metabolite. The amplitude ratios between the target transition and qualifier transition should be added into the Ratio (%) column. If the user doesn't want to use the ratio information, "–1" should be added as a dummy ratio. Also, MRMPROBS can accept one transition for one metabolite. Just set "100" for the metabolite in the Ratio (%) column.

- Decimal of precursor and product *m/z* is automatically converted to the integer value.

5.5.3 File Import and Create New Project (Figure 5.4)

1. Open "MRMPROBS.exe".
2. Create New project: File = > New project.

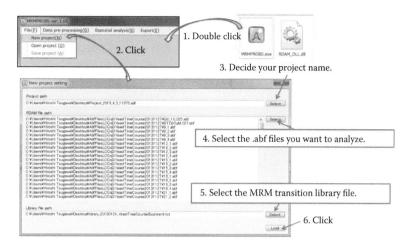

Figure 5.4 How to import the analysis files. (From Tsugawa et al., *Analytical Chemistry,* 2013, 85: 5191–5199.)

3. Select the project file path.
4. Select analysis files.
5. Select the MRM transitions library file path.
6. Click "Load" button.

5.5.4 Parameter Setting

Data Preprocessing => Parameter Setting

1. Smoothing: Select the required smoother from either the Savitzky–Golay smoother, linear weighted moving average (LWMA), or none. We recommend the LWMA for the smoother.
2. Smoothing level: The +/– number of data points set in the textbox is used for the smoother. We recommend a value from 3 to 10.
3. Base window (min): Add the average distance from the peak top to the peak edge in your chromatogram.
4. Area noise factor: Add the threshold of relative abundance from the highest peak intensity of a chromatogram in the same MRM transition.

5. Peak edge factor: Add the threshold of the peak edge intensity from the peak top. This is used to judge whether a detected peak is divided into two peaks or is merged as one peak.
6. RT tolerance (min) to reference: Add the retention time shift of detected peaks in your MRM data sets.
7. Click "Set" button.

5.5.5 Data Matrix Construction

Data preprocessing = > Raw data matrix create

5.5.6 Data Organization

In MRM-based metabolomics, the most important issue is to quantify the targeted metabolites accurately. In fact, MRMPROBS enables the users to quantify and identify the metabolites easily. However, we recommend that users confirm the result of peak identification and quantification manually. Modification of the results is sometimes required, especially when the user wants to employ the peak area for peak quantification.

MRMPROBS also provides a user-friendly graphical interface. This program supports three types of quantification, including peak height, peak area above zero value, and peak area above baseline. The user can select the objective option for peak quantification as required. MRMPROBS also provides many mouse-controlled functions. More information regarding the details of the mouse functions including click and drag is provided in the manual (Figure 5.5).

5.5.7 Statistical Analysis

Although MRMPROBS implements statistical analyses such as the normalization method, graph analysis, principal component analysis, and so on, details of the statistical analysis are provided in Chapter 6. The data matrix in MRMPROBS can be exported to both txt or csv format: Export -> txt or csv format.

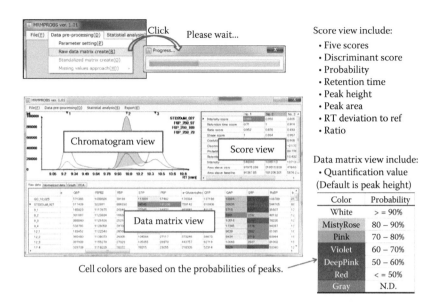

Figure 5.5 MRMPROBS. (From Tsugawa et al., *Analytical Chemistry,* 2013, 85: 5191–5199.)

References

1. Putri, S. P., Yamamoto, S., Tsugawa, H., and Fukusaki, E. Current metabolomics: Technological advances, *Journal of Bioscience and Bioengineering*, 2013, 116: 9–16.
2. Soga, T., Ueno, Y., Naraoka, H., Ohashi, Y., Tomita, M., and Nishioka, T.. Simultaneous determination of anionic intermediates for *Bacillus subtilis* metabolic pathways by capillary electrophoresis electrospray ionization mass spectrometry, *Analytical Chemistry*, 2002, 74: 2233–2239.
3. Soga, T., Igarashi, K., Ito, C., Mizobuchi, K., Zimmermann, H. P., and Tomita, M., Metabolomic profiling of anionic metabolites by capillary electrophoresis mass spectrometry, *Analytical Chemistry*, 2009, 81: 6165–6174.
4. Fiehn, O., Kopka, J., Dormann, P., Altmann, T., Trethewey, R. N., and Willmitzer. L., Metabolite profiling for plant functional genomics, *Nature Biotechnology,* 2000, 18: 1157–1161.

5. Kopka, J., Current challenges and developments in GC-MS based metabolite profiling technology, *Journal of Biotechnology,* 2006, 124: 312–322.

6. Lisec, J., Schauer, N., Kopka, J., Willmitzer, L., and Fernie, A. R., Gas chromatography mass spectrometry-based metabolite profiling in plants, *Nature Protocols,* 2006, 1: 387–396.

7. Bottcher, C., Roepenack-Lahaye, E. V., Schmidt, J., Schmotz, C., Neumann, S., Scheel, D., and Clemens, S., Metabolome analysis of biosynthetic mutants reveals diversity of metabolic changes and allows identification of a large number of new compounds in *Arabidopsis thaliana*, *Plant Physiology,* 2008, 147: 2107–2120.

8. Matsuda, F., Yonekura-Sakakibara, K., Niida, R., Kuromori, T., Shinozaki, K., and Saito, K., MS/MS spectral tag (MS2T)-based annotation of non-targeted profile of plant secondary metabolites, *The Plant Journal*, 2009, 57: 555–577.

9. Soga, T., Baran, R., Suematsu, M., Ueno, Y., Ikeda, S., Sakurakawa, T., Kakazu, Y., Ishikawa, T., Robert, M., and Nishioka, T., et al., Differential metabolomics reveals ophthalmic acid as an oxidative stress biomarker indicating hepatic glutathione consumption, *Journal of Biological Chemistry,* 2006, 281: 16768–16776.

10. Kind, T. and Fiehn, O., Advances in structure elucidation of small molecules using mass spectrometry, *Bioanalytical Reviews,* 2010, 2: 23–60.

11. Saito, K. and Matsuda, F., Metabolomics for functional genomics, systems biology, and biotechnology, *Annual Review of Plant Biology,* 2010, 61: 463–489.

12. Matsuda, F., Shinbo, Y., Oikawa, A., Hira, M. Y., Fiehn, O., Kanaya, S., and Saito, K., Assessment of metabolome annotation quality: A method for evaluating the false discovery rate of elemental composition searches, *PloS ONE,* 2009, 4: e7490.

13. Neumann, S. and Bocker, S., Computational mass spectrometry for metabolomics: Identification of metabolites and small molecules, *Analytical and Bioanalytical Chemistry,* 2010, 398: 2779–2788.

14. Putri, S. P., Nakayama, Y., Matsuda, F., Uchikata, T., Kobayashi, S., Matsubara, A., and Fukusaki, E., Current metabolomics: Practical applications, *Journal of Bioscience and Bioengineering,* 2013, 115: 579–589.

15. Tsugawa, H., Tsujimoto, Y., Sugitate, K., Sakui, N., Nishiumi, S., Bamba, T., and Fukusaki, E., Highly sensitive and selective analysis of widely targeted metabolomics using gas chromatography/triple-quadrupole mass spectrometry, *Journal of Bioscience and Bioengineering,* 2012, 117: 122–128.

16. Voehringer, P., Fuertig, R., and Ferger, B., A novel liquid chromatography/tandem mass spectrometry method for the quantification of glycine as biomarker in brain microdialysis and cerebrospinal fluid samples within 5 min, *Journal of Chromatography,* 2013, 939: 92–97.

17. Sawada, Y., Akiyama, K., Sakata, A., Kuwahara, A., Otsuki, H., Sakurai, T., Saito, K., and Hirai, M.Y., Widely targeted metabolomics based on large-scale MS/MS data for elucidating metabolite accumulation patterns in plants, *Plant and Cell Physiology,* 2009, 50: 37–47.

18. Sawada, Y. and Hirai, M. Y., Integrated LC-MS/MS system for plant metabolomics, *Computational and Structural Biotechnology Journal,* 2013, 45: e201301011.

19. Kato, H., Izumi, Y., Hasunuma, T., Matsuda, F., and Kondo, A., Widely targeted metabolic profiling analysis of yeast central metabolites, *Journal of Bioscience and Bioengineering,* 2012, 113: 665–673.

20. Luo, B., Groenke, K., Takors, R., Wandrey, C., and Oldiges, M., Simultaneous determination of multiple intracellular metabolites in glycolysis, pentose phosphate pathway and tricarboxylic acid cycle by liquid chromatography-mass spectrometry, *Journal of Chromatography A,* 2007, 1147: 153–164.

21. Bajad, S. U., Lu, W., Kimball, E. H., Yuan, J., Peterson, C., and Rabinowitz, J. D., Separation and quantitation of water soluble cellular metabolites by hydrophilic interaction chromatography-tandem mass spectrometry, *Journal of Chromatography A,* 2006, 1125: 76–88.

22. Sawada, Y., Nakabayashi, R., Yamada, Y., Suzuki, M., Sato, M., Sakata, A., Akiyama, K., Sakurai, T., Matsuda, F., Aoki, T., et al., RIKEN tandem mass spectral database (ReSpect) for phytochemicals: A plant-specific MS/MS-based data resource and database, *Phytochemistry,* 2012, 82: 38–45.

23. Tsugawa, H., Arita, M., Kanazawa, M., Ogiwara, A., Bamba, T., and Fukusaki, E., MRMPROBS: A data assessment and metabolite identification tool for large-scale multiple reaction monitoring based widely targeted metabolomics, *Analytical Chemistry,* 2013, 85: 5191–5199.

6
Statistical Analysis

Chapter 6
Statistical Analysis

Hiroshi Tsugawa and Takeshi Bamba

Chapter Outline

6.1 Introduction

A data matrix (sample vs. metabolites) acquired from a mass spectrometry-based metabolomics study includes not only biological information but also technical noise. Thus, meaningful information has to be extracted from the complicated and huge data matrix by means of statistical analysis. It is very important to use and interpret the statistical result accurately. Otherwise, critical misunderstanding of the data from a biological point of view might occur. The purpose of this chapter is to aid researchers in finding novel biological insights accurately from statistical analyses.

There are numerous materials available for learning the mathematical details of univariate or multivariate statistical analysis, therefore we focus on how to do the multivariate and univariate analysis and interpret the statistical results. The summary of statistical analysis methods introduced here is described in Table 6.1.

The step-by-step procedure is described in this chapter. First, we introduce an Excel-based statistical analysis method which is a freely available macro file developed by Dr. Tsugawa for multi t test,[1,2] graph analysis, principal component analysis (PCA),[3] correlation analysis,[4] projection to latent structure regression (PLS-R),[5] and projection to latent

TABLE 6.1
Summary of Statistical Analyses Described in This Chapter

Statistical Method	Usage
Principal component analysis (PCA)	· Make a summary review of samples and metabolites in the informative axis on the basis of the data variance.
	· Find the outlier samples.
Hierachical cluster analysis (HCA)	· See the similarities (distances) among samples and metabolites on the multidimensional spaces.
Correlation analysis (CA)	· See the correlation relationship of the metabolites especially via time-course experiment.
Projection to latent structure regression (PLS-R)	· Find the sample differences and significant metabolites with the specific and continuous Y variables.
	· Construct the robust regression model by the latent variables.
Projection to latent structure discriminant analysis (PLS-DA)	· Find the sample differences and significant metabolites with the binary (0, 1) values.
	· Construct the robust discriminant model by the latent variables.
T test with Bonferroni correction	· Define the significant value with the aiming of reducing the false-positive discoveries.
T test with false discovery rate correction	· Define the significant value with the false-positive rate correction.

Source: Reprinted with permission from Dr. Hiroshi Tsugawa.

structure discriminant analysis (PLS-DA).[6] Because this program is implemented in Microsoft Excel (Visual Basic for application), it is expected to be easily operated by a beginner. Unfortunately, because HCA is not implemented in this macro program, we introduce the easy-to-use heat map analysis by hierarchical clustering analysis (HCA)[7] in statistical language R instead.

6.2 General Aspect of Multivariate and Univariate Analysis

6.2.1 Data Pretreatment

Before multivariate analysis, a data pretreatment method (i.e., centering, auto scaling, Pareto scaling, range scaling, vast scaling, log transformation, and power transformation) should be performed first in order to get meaningful statistical results. More important, results from multivariable analysis should be interpreted according to the data pretreatment method. These different pretreatment methods emphasize different aspects of the data and each has its own merits and drawbacks. The choice of the pretreatment method depends on the biological question that needs to be answered, properties of the data set, and the data analysis method selected. The details of the data pretreatment methods were given in a previous report.[7] The summary of data pretreatment methods introduced here is described in Table 6.2.

6.2.2 Principal Component Analysis

In the study of metabolomics, principal component analysis (PCA) is the most frequently used method for data mining as a nonsupervised approach.[3] In multivariate analysis, each sample is regarded as a point in the multivariate dimension. To illustrate, consider 100 samples with 500 metabolites' information in a data matrix. In this case, there are 100 points in 500 dimensional spaces. PCA finds a new meaningful axis based on the variance of 100 points in 500 dimensional

TABLE 6.2
Summary of Data Pretreatment Methods Described in this Chapter

Class	Method	Formula	Features on Multivariate Analysis
Centering	Centering	$\widetilde{x}_{ij} = x_{ij} - \overline{x}_i$	· The result is mainly based on high magnitude variables. · Low sensitivity to noise.
Scaling	Auto scaling	$\widetilde{x}_{ij} = \dfrac{x_{ij} - \overline{x}_i}{s_i}$	· All variables are equally treated. · High sensitivity to noise.
	Pareto scaling	$\widetilde{x}_{ij} = \dfrac{x_{ij} - \overline{x}_i}{\sqrt{s_i}}$	· The original measurement scale is relatively kept compared to auto scaling. · Low sensitivity to noise compared to auto scaling. · The tendency of statistical result is close to centering method rather than auto scaling.
Transformation	Log transformation	$\widetilde{x}_{ij} = \log_{10} x_{ij}$	· Heteroscedasticity is corrected. · Pseudo scaling and similar to auto scaling. · Zero value is not acceptable.
	Power transformation	$\widetilde{x}_{ij} = \sqrt[p]{x_{ij}}$	· Heteroscedasticity is corrected. · Pseudo scaling and similar to auto scaling. · Zero value is acceptable.

Here, the number of samples and metabolites is n and m, respectively. $\overline{x_i}$ and s_i are based on the following equation:

$$\overline{x_i} = \sum_{j=1}^{n} x_{ij}$$

$$s_i = \sum_{j=1}^{n} \sqrt{\frac{(x_{ij} - \overline{x_i})^2}{j-1}}$$

$\overline{x_i}$ and s_i means the average and standard deviation of focused metabolite, respectively. In the power transformation, p is frequently set to 4.

Source: Reprinted with permission from Dr. Hiroshi Tsugawa.

TABLE 6.3
Terminology Used in Principal Component Analysis

Term	Usage
Score	Inner product between x value and its coefficient.
	The plot is used to understand the features of samples.
Loading	Value projecting x value to the constructed axis.
	The plot is used to see the variable importance constructing the informative axis.
Contribution ratio	Amount of information explained by the principal component.

Source: Reprinted with permission from Dr. Hiroshi Tsugawa.

spaces. By projecting 100 points to the meaningful axis (principal component: PC), the main differences among samples can be seen. A list of common terminology in PCA can be seen in Table 6.3.

PCA divides a data matrix into three parts on the basis of the variance within the data: the score matrix, the loading matrix, and the residual matrix. Researchers can identify the features of their biological samples from the score matrix. For searching sample differences 2D or 3D score plots are generally used. Moreover, researchers can identify the metabolites contributing to the differences among samples from the loading matrix. By comparing the score plots and loading plots, the relationship between the samples and metabolite features can be understood.

6.2.3 Hierarchical Cluster Analysis

Hierarchical cluster analysis (HCA) is another frequently used method in metabolomics study for the nonsupervised approach.[7] Metabolite information for each sample is expressed by a vector; that is, it is represented by a point in the multidimensional space. The distances between these points are calculated, and the clustering procedure depends on the distances within the multidimensional space. There are many distance calculation methods and clustering procedures but in our case, the Euclidean distance is most frequently used for distance calculation. The correlation coefficient as the distance approach is also used especially for the time-course data set. In the case of clustering algorithms, the frequently used methods are single, complete, average, centroid, and ward linkage. However, the optimal clustering procedure method is dependent on the data set.

The advantage of HCA is that researchers can interpret the biological features in two-dimensional visualization spaces. For analysis of a large number of samples (100–1,000 samples), PCA occasionally makes data interpretation difficult. Therefore, HCA enables us to understand easily the summary or the feature of multivariate samples. Recently, a number of statistical tools have been used to apply HCA to both the sample and variable apexes to visualize the features in one figure.

6.2.4 Correlation Analysis

In cases where researchers want to see the relationship of metabolites, the covariation of metabolites (e.g., from high grade to low grade, from start time to end time, from older to younger, etc.) should be examined to see the metabolite relationship in the living organisms.[8] For this purpose, the correlation coefficient is frequently used to see the metabolite relationship.

6.2.5 Projection to Latent Structure Regression and Discriminant Analysis

In addition to the nonbias approach such as PCA, HCA, and correlation analysis, the supervised multivariate technique is also used to identify interesting metabolites. The PLS algorithm is frequently used in the metabolomics research field to see the significant metabolites.[5,6] The features of the PLS algorithm are the following:

- The Y-matrix (or dummy variable) is used to find the new dimensional axis.
- The Y-correlated information is extracted from the X matrix.

- The PLS algorithm works well on a larger number of variables (metabolites) compared to the number of the object (samples).
- The significant metabolites related to Y variables are easily extracted.

The terminologies described in PLS-based multivariate analysis are described in Table 6.4.

The orthogonal PLS algorithm is also useful because particularly in the univariate Y, the OPLS algorithm can remove the uncorrelated data from the X to Y matrix, where the number of the latent variables correlated to Y is generally one.[9] Because of this feature, users can focus on the first component only. However, the PLS and OPLS algorithms have completely the same performance. Moreover, especially for the beginner, the PLS algorithm should be understood first to use such PLS-based algorithms accurately. Here, we describe the use of the PLS algorithm in Microsoft Excel and the interpretation of the result. When the Y matrix has specific variables, such as the food quality evaluated by a sensory test, the PLS regression is a convenient method to extract the significant metabolites correlated with Y variables.

On the other hand, if the Y matrix does not have specific variables but has some biases, PLS-DA is frequently used.[5] Instead of specific variables, the discriminant approach utilizes binary vectors consisting of 0 and 1. Also, the OPLS algorithm instead of the PLS is useful to interpret the result easily

TABLE 6.4
Terminologies Used in PLS-Based Multivariate Analysis

Term	Usage
Q2	Guidepost to determine the optimal number of latent variables
VIP	Criteria for selecting important variables of the PLS model
Coefficient	Coefficient value of X for Y regression, which is used to see the variable importance
RMSEE	Index from internal validation to evaluate the accuracy and precision of the model
RMSEP	Index from external validation to evaluate model robustness

Source: Reprinted with permission from Dr. Hiroshi Tsugawa.

and can use the S-plot to find significant metabolites.[10,11] The PLS-DA procedure is mostly the same with PLS-R but the purpose is different between the two. PLS-DA is a method to distinguish two groups and to find the important metabolites. Using three or more discriminant analysis procedures is not recommended because it makes interpretation of the statistical result difficult.

6.2.6 Univariate Analysis

Univariate analyses such as the T test, U test, and ANOVA are also important for data mining. Compared with multivariable analysis, univariate analysis offers a simple and instinctive result. However, because the metabolome data sets include a large number of variables, the significance level should be appropriately determined to reduce the number of false positives and false negatives. Familywise error rate correction, such as a Bonferroni correction,[1] is a conservative approach used to reduce false positives. In contrast, FDR (false discovery rate) correction[2] is a highly sensitive method of reducing false negatives, and has recently gained popularity in metabolomics.

Moreover, it is more instinctive to see the bar graph or line chart of each metabolite because each metabolite can be considered for getting new insights with biological significance. However, subjective interpretation should not be performed without a statistical test such as the T test.

6.3 Step-by-Step Procedure of Multivariate and Univariate Analysis

Here, we demonstrate the protocol for statistical analysis as well as the interpretation of the results using the Japanese green tea data set. We recommend that readers avoid skipping any section of this chapter and follow the step-by-step procedure and its interpretation. First, the Excel macro program is demonstrated.

Software

- Tool for statistical analysis on Microsoft Excel: http://prime.psc.riken.jp/(Please follow a link: "Targeted and Non-targeted analysis software" = > "Tool for statistical analysis on Microsoft Excel")

Environment

- Windows OS
- Microsoft Excel 2007 or later

Available Method

- Normalization method by internal standard
- Tool for generation of bar graph
- Tool for generation of line chart
- Multi T test
- PCA
- Correlation analysis
- PLS-R
- PLS-DA

Note:: The above functions are also implemented in the AIoutput program described in Chapter 4.

Example Data

- The file "GreenTeaMetabolome.csv" in the included downloaded folder is used here as sample data.
- Samples are Japanese green tea ranked in an agricultural fair. The sample name shows the ranking number, that is, 1, 6, 11, 16, 21, 31, 36, 41, 46, 51 ($n = 3$). Here, in order to easily explain the statistical analysis, the class indexes are arranged as follows: No. 1–6, No. 11–36, and No. 41–51 are classified as 1, 2, and 3, respectively. Class 1, Class 2, and Class 3 are subjectively considered as high grade, middle grade, and low grade from the agricultural fair. In total, the number of samples, metabolites, and classes is 30, 225, and 3, respectively.

- Data file must be saved in CSV file format.
- The detail of the required format is written in the manual found in the downloaded folder. The first row corresponds to the class index, the second row corresponds to the sample name, and the later rows include the compound name and its intensity variables. An explanation of the contents format of an Excel file can be seen in Figure 6.1.

(1) File import

   a. Open "StatisticalAnalysis.xlsm" file.

   b. Click "Statistical Analysis" button in "Base data" sheet.

   c. Click "File Import" and select the analysis file (In this case, GreenTeaMetabolome.csv). Please see Figure 6.2.

Note: If you get the error message "Compile Error: Can't find project or library", see part 3 of the data processing section in Chapter 4.

(2) Class rearrangement

- Users can rearrange the class indexes of samples by means of "Class change" button as shown below (Figure 6.3).

A	B	C	D	E	F	G	H	I	J	K	L	M	N	O
Class	1	1	1	2	2	2	3	3	3	4	4	4	5	
Row Label: Column: Sample	No1-1	No1-2	No1-3	No6-1	No6-2	No6-3	No11-1	No11-2	No11-3	No16-1	No16-2	No16-3	No21-1	No21-
2-Hydroxypyridine: C02502	3736	2722	2391	3438	2147	2407	5118	3753	4402	4245	3899	3588	4383	34
Pyruvate+Oxaloacetic acid: C00022+C0	251	156	172	234	163	181	211	184	190	198	181	183	194	
Lactic acid: C00186	557	234	260	425	169	200	251	193	206	151	148	149	148	
Glycolic acid: C00160	323	231	276	182	146	162	215	189	209	205	207	202	131	
Alanine_2TMS_Major: C00041	16355	15239	14263	20288	15808	17903	12653	13560	11157	13821	11732	11832	8748	92
C11_Alkane	499	475	482	609	353	418	461	351	406	447	430	366	418	
n-Butylamine: C18706	1390	1335	1456	1242	1157	1282	1207	993	1364	1244	1320	1127	1097	
Oxalate: C00209	109056	79250	83403	19787	14424	14021	25524	25973	26964	43401	36423	31398	15205	141
3-Hydroxybutyrate: C01089	681	522	401	726	553	526	658	462	746	1109	1116	887	847	
2-Aminobutyric acid: C02251	180	147	149	181	141	172	96	98	88	154	122	138	75	
Malonic acid: C00383	2008	1389	1341	1349	978	935	1298	1071	952	1499	1182	1078	1146	
Valine_2TMS_Major: C00183	1991	1653	1538	4324	3327	3417	1396	1386	1257	1623	1390	1421	1244	
Unknown_42	2528	2035	1782	1671	1188	1210	2459	2182	2494	2338	2047	1985	1607	
Unknown_45	2267	2481	3266	2188	2151	2393	1771	2108	2644	2101	2049	2193	2022	
Urea: C00086	225	169	185	2	2	2	176	153	162	223	206	200	407	
Serine_2TMS_Minor: C00065	5002	4259	4637	3755	3528	4336	3691	2615	4376	2210	3142	2235	3022	2
2-Aminoethanol: C00189	1187	1167	1279	1667	1461	1465	725	806	803	1216	1248	1171	1008	
Unknown_58	20113	18560	19692	20936	16908	19085	10712	10790	11478	16077	16077	15437	9587	99
Phosphate: C00009	76942	70626	74834	79784	64091	72110	40373	40844	43067	61392	61136	58624	35877	374
Leucine_2TMS: C00123	1197	1291	1216	3985	3635	3443	795	922	772	1359	1162	1191	1097	11
Isoleucine_2TMS: C00407	857	919	850	2379	2097	2058	840	922	775	1172	1032	1047	733	
Proline: C00148	2155	2411	2224	3502	3177	2968	1215	1496	1139	1749	1377	1502	1438	15
Maleic acid: C01384	2	2	3	2	2	2	113	209	2	114	176	64		
Glycine_3TMS_Major: C00037	1269	1102	1112	1388	1178	1355	861	834	839	818	785	763	789	
Succinic acid(or aldehyde): C00042	18422	16495	16589	12101	10329	11123	12149	11815	12142	12776	12569	11686	9992	100
Glyceric acid: C00258	1719	1529	1612	1344	1156	1216	2555	2564	2584	2981	2854	2729	2034	20
Fumaric acid: C00122	728	665	653	704	637	696	591	588	604	617	617	593	546	
Serine_3TMS_Major: C00065	12226	14159	14118	11054	10360	11490	7814	9620	7993	9488	8585	9231	7734	84

Figure 6.1 Explanation of the contents format of an Excel file. (Reprinted with permission from Dr. Hiroshi Tsugawa.)

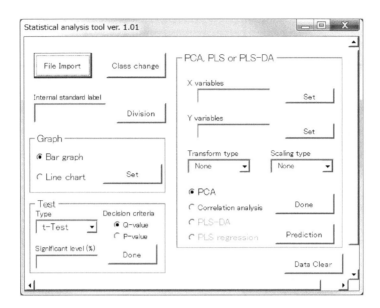

Figure 6.2 Screen shot of the Excel macro's form. (Reprinted with permission from Dr. Hiroshi Tsugawa.)

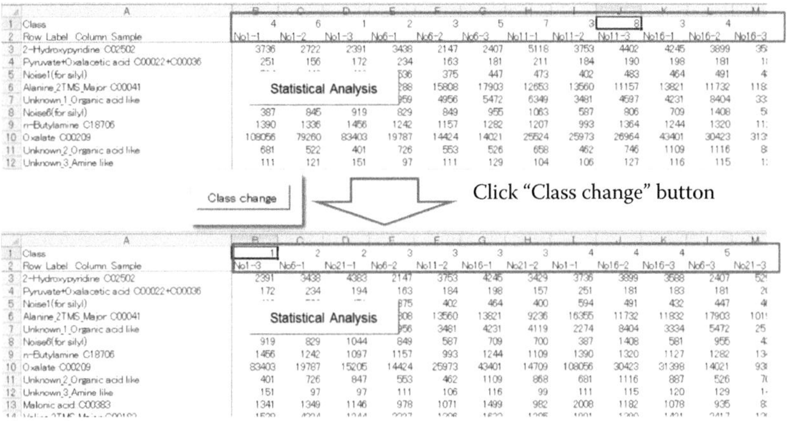

Figure 6.3 How to set the class ID. (Reprinted with permission from Dr. Hiroshi Tsugawa.)

	A	B	C	D	E	F	G	H	I	J	K	L	M
1	Class	1	1	1	2	2	2	3	3	3	4	4	
2	Row Label: Column: Sample	No1-1	No1-2	No1-3	No6-1	No6-2	No6-3	No11-1	No11-2	No11-3	No16-1	No16-2	No16-3
81	Unknown_29_Organic acid like	212	193	176	364	315	346	187	191	207	172	152	1
82	Arabitol C01904	297	257	278	219	190	226	162	156	172	211	198	2
83	Unknown_30_Organic acid like	360	323	360	299	243	276	212	204	234	283	249	3
84	Unknown 31	573	530	569	526	513	502	475	514	536	545	501	5
85	Ribitol C00474	39765	34621	37353	35062	33284	32135	32075	33416	33906	36190	33103	323
86	Unknown_32	1536	1330	1432	1364	1298	1232	1239	1274	1287	1398	1275	12
87	Unknown_33	264	222	246	234	233	212	212	229	211	239	223	2
88	Unknown_34_Sugar like	2858	2530	2838	2962	2666	2935	953	925	1003	2203	2054	22
89	Putrescine C00134	430	420	456	462	441	509	316	306	351	266	245	2

Internal standard label | Ribitol C00474 | Division ⟹ Click "Division" button

	A	B	C	D	E	F	G	H	I	J	K	L	M
1	Class	1	1	1	2	2	2	3	3	3	4	4	
2	Row Label: Column: Sample	No1-1	No1-2	No1-3	No6-1	No6-2	No6-3	No11-1	No11-2	No11-3	No16-1	No16-2	No16-
81	Unknown_29_Organic acid like	0.005469	0.005575	0.004712	0.010382	0.009464	0.010767	0.00583	0.005716	0.006105	0.004753	0.004592	0.005-
82	Arabitol C01904	0.007662	0.007423	0.007443	0.006246	0.005708	0.007033	0.005061	0.004668	0.005073	0.00583	0.005981	0.00
83	Unknown_30_Organic acid like	0.009287	0.00933	0.009638	0.008528	0.007301	0.008589	0.00661	0.006105	0.006901	0.00782	0.007522	0.009
84	Unknown 31	0.014781	0.015309	0.015233	0.015002	0.015413	0.015622	0.014809	0.015382	0.015806	0.015059	0.015135	0.015
85	Ribitol C00474	1	1	1	1	1	1	1	1	1	1	1	1
86	Unknown_32	0.039623	0.038416	0.038337	0.038903	0.038697	0.038338	0.038628	0.038125	0.037958	0.038629	0.038516	0.037
87	Unknown_33	0.00681	0.006412	0.006586	0.006674	0.007	0.006597	0.00661	0.006853	0.006223	0.006604	0.006737	0.006
88	Unknown_34_Sugar like	0.073726	0.073077	0.075978	0.084479	0.080099	0.091333	0.029712	0.027681	0.029582	0.060873	0.062049	0.070
89	Putrescine C00134	0.011092	0.012131	0.012208	0.013177	0.01325	0.015839	0.009852	0.009157	0.010352	0.00735	0.007401	0.008

Figure 6.4 How to do the normalization by an internal standard. (Reprinted with permission from Dr. Hiroshi Tsugawa.)

(3) Normalization of the intensity variables by the internal standard's variables
   a. Copy and paste or enter the label information of internal standard.
   b. Click "Division" button. See Figure 6.4.

Note: If the internal standard is added to the samples (Chapter 4, GC/MS: ribitol; Chapter 5, LC/QqQ/MS: PIPES), the intensity variables of metabolites should be normalized by the internal standard's intensity to correct the accident error of sample extraction and instrument analysis. For the normalization method, some important and useful methods have been reported.[12,13]

(4) Making the bar graph and line chart
   • Bar graph: for comparison among groups (Figure 6.5).
      a. Check "Bar graph" in Graph section and click "Set" button.
      b. Select the error bar criteria from standard deviation or standard error.
      c. Click "Done" button.
   • Line chart: for comparison of time course data among groups. Because the demonstrated data set is not time course data, the procedure for making a line chart is explained here (Figure 6.6).

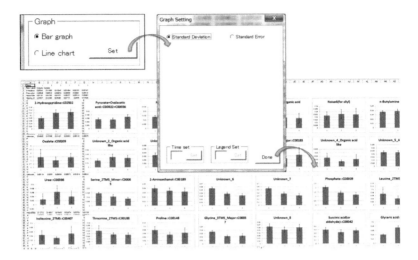

Figure 6.5 How to make the bar graphs of metabolites. (Reprinted with permission from Dr. Hiroshi Tsugawa.)

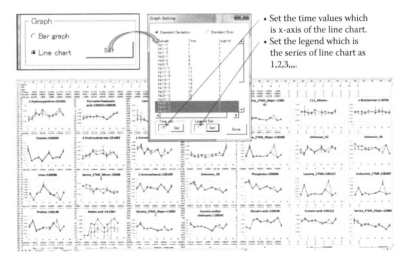

Figure 6.6 How to make the line charts of metabolites. (Reprinted with permission from Dr. Hiroshi Tsugawa.)

1. Check "Line chart" in Graph section and click "Set" button.
2. Select the error bar criteria from standard deviation or standard error.
3. Set the time information as the *x*-axis using the following procedures:
   a. Drag the rows.
   b. Add the time value in the textbox of "Time set".
   c. Click "Set" button.
4. Set the legend information for defining the groups by using the same procedures described above.
5. Click "Done" button.

(5) *T* test

- Here, we introduce how to do the *T*-test with two types of methods, that is, Bonferroni correction (1) and FDR (2) (false discovery rate) correction. Please see Figure 6.7 for an illustration on *T*-test.
- Bonferroni correction: (for reducing false positives)
  a. Check "*P*-value" for the decision criteria.
  b. Calculate the significant level. When setting the 5% significance, "add 5/n" should be added to the textbox

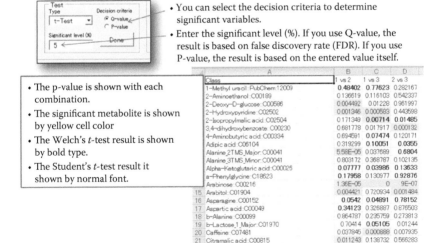

Figure 6.7 How to do the *T*-test. (Reprinted with permission from Dr. Hiroshi Tsugawa.)

of significant level. Here, "n" means the number of metabolites. In this case, n is 225. Therefore, add 0.022 to the textbox.

  c. Click "Done".

- FDR correction: (for reducing false negatives. Nowadays, it has been frequently used in metabolomics study).

  a. Check "Q-value" for the decision criteria.

  b. Add 5 to the textbox of significant level. The "5" in FDR correction means that 5% false positives would be included in the significant metabolites.

  c. Click "Done".

(6) Principal component analysis

- Here, we explain how to do the PCA and interpret the score and loading plots of the PCA result (Figure 6.8).

  a. Select the labels as X variables by dragging with a mouse.

  b. Click "Set button".

  c. Select the "Transform type" and "Scaling type". Here, "None" and "Auto scale" are selected for the data transforming and scaling method, respectively. This option means that the averages and standard deviations of all variables of each metabolite, that

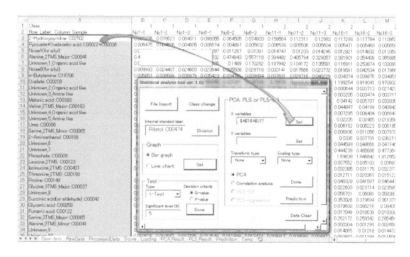

Figure 6.8 How to set the X-variables for multivariate analysis. (Reprinted with permission from Dr. Hiroshi Tsugawa.)

is, each row, are standardized to 0 and 1, respectively. Because the amplitude range of metabolites is 10^4–10^6 orders, the extracted PCA axis is mainly based on high-amplitude metabolites. Therefore, the auto scale is frequently used to deal with the metabolite information as equal value. Note that there is a disadvantage in the auto scaling method in that noise values are also increased. The details of other transformation and scaling methods have been previously reported.[14]

d. Check "PCA" radio button and click "Done". The result of a PCA result can be seen in Figure 6.9.

e. First, find the outlier samples in the PCA score plot. As shown above, the x-axis and y-axis show the first principal component (PC1) and the second principal component (PC2), respectively. In this example, there is no outlier and each rank sample of Japanese green tea is closely clustered. This indicates that the experimental procedures such as the sample extraction or instrument analysis were successfully done. On the other hand, when outliers are present, the raw chromatogram of outlier samples should be checked again. Sometimes, there are cases that the sample injection of the auto sampler is not successfully performed. After consideration for such outliers, reanalysis without them should be performed.

f. Interpret the PCA score plot. The interpretation for each extracted axis of PCA should be "subjectively"

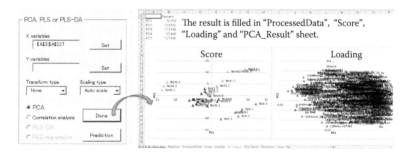

Figure 6.9 Principal component result. (Reprinted with permission from Dr. Hiroshi Tsugawa.)

given; that is, interpretation of the PCA score plot depends on the researchers themselves. In addition, the interpretation should be given to each axis independently because the principal component is mathematically orthogonal to each. In this example, the second axis of the PCA score plot seems related to the green tea quality: high-, middle-, and low-grade tea. Therefore, we subjectively decided that the second axis represents Japanese green tea's quality.

g. Interpret the PCA loading plot. The disadvantage of using auto scaling is that it's hard to see the 2D loading plot as shown above. Therefore, we made a bar graph of each PC axis by using the loading value resulting in the "Loading" sheet as shown in Figure 6.10. Note that the loadings in each axis are sorted by the values. The x-axis and y-axis show the compound name and loading value, respectively. Here, we show a significant criterion for selection of the significant metabolites although the method has not been well known in the metabolomics research field. (The contents of Step h are for advanced learners and are not applicable by default setting in this Excel macro. Novice learners can skip Step h.)

h. The loading value (p) is defined as the correlation coefficient between the score value (t) and the raw data (X) as can be seen in Equation 6.1.[15]

$$p(\text{corr}) = \frac{X^T t}{||X|| \, ||t||} \tag{6.1}$$

However, this Excel macro and AIoutput don't generate the coefficient but output the covariance standardized by the score scalar between the score value and raw data as follows (Equation 6.2):

$$p = \frac{X^T t}{||t||^2} \tag{6.2}$$

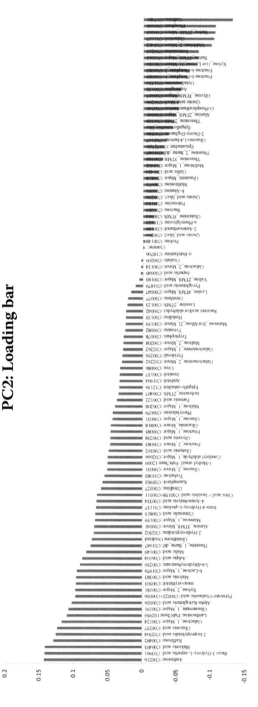

Figure 6.10 Loading bars in principal component analysis. X axis and Y axis show the compound name and its loading value, respectively. (Reprinted with permission from Dr. Hiroshi Tsugawa.)

The commercial software, SIMCA-P and Pirouette, also generate this loading value by default. We always generate the $p(\text{corr})$ by means of Excel calculation of the data stored in "ProceededData", "Score", and "Loading" sheets. If you want to use it as the default setting, rewrite the macro code as follows:

Development tab = > Visual Basic Editor = > "MultiVariableAnalysis" module = > "PCAanalysis()" procedure = > go to the end of this procedure and change the blue code to red code as shown in Equation 6.3. Then the result is generated in the "Loading" sheet.

```
xyz:
ReDim PC(n, k)
For i = 2 To n
  For j = 3 To k
    PC(i, j) = tnew(j) * p(i)
    a(i, j) = a(i, j) -PC(i, j)
  Next j
Next i

For i = 3 To k
  WS2.Cells(i - 1, q + 1) = tnew(i)
Next i

temp_t = 0
For i = 3 To k
  temp_t = temp_t + tnew(i) * tnew(i)
Next i

For i = 2 To n
  temp_pl = 0
  For j = 3 To k
    temp_pl = temp_pl + E(i, j) * E(i, j)
  Next j
  WS3.Cells(i, q + 1) = p(i) * Sqr(temp_t) / Sqr(temp_pl)
Next i

For i = 2 To n
  WS3.Cells(i, q + 1) = p(i)
Next i
```

$$(6.3)$$

For the statistical T test of the correlation coefficient, the t value is calculated as follows (Equation 6.4):

$$t = \frac{r\sqrt{n-2}}{\sqrt{1-r^2}}$$ (6.4)

We utilize the fact that this t value follows the t distribution with $n - 2$ degrees of freedom where, n means the number of samples and r is the absolute value of the correlation coefficient. In this example, the p(corr) of PC2 loading was calculated because we saw an interesting pattern related to the green tea quality in the PC2 axis. The t-value was calculated by using the correlation coefficient as shown above. The p value was calculated by using the "tdist" function in Microsoft Excel. Finally, the p value was corrected by the Bonferroni method. Here, because the number of metabolites (i.e., the number of statistical tests) is 225, the expectation value of false positives is 225 * 0.05 = 11.25. To avoid this misunderstanding, we utilized the Bonferroni method frequently used for correction of the familywise error rate (FWER). The final p value is the product of the p value and the metabolite number (225). Then, we recognized the final value less than 0.05 as the significant metabolites. The significant value is shown in Table 6.5. Here, we just show the identified metabolites.

    i. Interpret the above result

By means of the above method, the 11 metabolites (i.e., UDP-D-Glucuronate, Arabinose, Threo-b-HydroxyAspartic acid, Shikimic acid, Caffeine, Raffinose, 2-Isopropylmalic acid, Glucono-1,5-lactone_2, Galactose_1, Lanthionine, and Gluconic acid) were objectively considered as the significant metabolites related to green tea quality. By focusing on the metabolites, the important metabolites or conventions are investigated for their relationship with food quality.

(7) Correlation analysis

The correlation coefficient is calculated for each metabolite. The significance of their relationship is examined in the

TABLE 6.5
Result of Significant Test for PCA Loadings

Compound_Name	t Value of PC2 Loading	p Value	Metabolite Number*p Value
UDP-D-Glucuronate	8.49514721	3.09211E-09	7.05001E-07
Arabinose	8.091703995	8.2457E-09	1.88002E-06
Threo-b-HydroxyAspartic acid	8.003419249	1.02488E-08	2.33672E-06
Shikimic acid	7.79956938	1.69986E-08	3.87568E-06
Caffeine	6.364113823	6.90953E-07	0.000157537
Raffinose	6.111437755	1.3567E-06	0.000309327
2-Isopropylmalic acid	5.941360095	2.14305E-06	0.000488616
Glucono-1,5-lactone_2	5.536786575	6.40952E-06	0.00146137
Galactose_1	4.886399404	3.78361E-05	0.008626622
Lanthionine	4.592621376	8.44152E-05	0.019246664
Gluconic acid	4.324145734	0.000175245	0.039955811
Pyrophosphate	4.143963262	0.000285312	0.065051195
Serine(2TMS)	4.122499149	0.000302309	0.068926363
Alpha-Ketoglutaric acid	4.074385854	0.000344117	0.07845873
Galactitol	3.999851747	0.000420374	0.095845254
meso-erythritol	3.793996736	0.000728	0.165984072
Melibiose_2	3.756485885	0.00080409	0.183332482
Oxalacetic acid	3.711341082	0.000906012	0.206570848
Malonic acid	3.708500232	0.000912832	0.208125676
b-Lactose	3.645463162	0.001077704	0.245716445
Xylose_2	3.602987759	0.001204799	0.274694119
3,4-dihydroxybenzoate	3.305494321	0.002603622	0.593625736
Adipic acid	3.252065322	0.002983742	0.680293195
Malic acid	2.839820405	0.008315433	1.89591864
Homoserine	2.793159549	0.009307189	2.122039147
Serine(3TMS)	2.763727124	0.009988871	2.277462631
Xylose_1	2.719689212	0.011096824	2.530075836
Theanine_1	2.630864707	0.013689942	3.121306688
Glutathione Oxidized	2.557809945	0.016234318	3.701424393
Indoxyl Sulfate	2.489616364	0.018998022	4.331548915

TABLE 6.5 (CONTINUED)
Result of Significant Test for PCA Loadings

Compound_Name	t Value of PC2 Loading	p Value	Metabolite Number*p Value
Alanine(3TMS)	2.468954525	0.019917345	4.541154593
Citramalic acid	2.410774666	0.022730135	5.182470846
2-Hydroxypyridine	2.34631398	0.026267994	5.989102653
4-Aminobutyric acid	2.316162308	0.028089231	6.404344627
Citric acid	2.303359216	0.028896697	6.588446817
Mannose_1	2.268282547	0.031217885	7.117677716
Hydroxyproline	2.250991014	0.032423106	7.392468135
Fructose-6-Phosphate_2	2.224087773	0.034381923	7.839078331
Fructose-6-Phosphate_1	2.090002545	0.045820372	10.44704487
Citrulline	2.070064605	0.047783491	10.89463602
Kaempferol	1.920953341	0.064973035	14.81385199
Trehalose	1.886459787	0.069643711	15.8787662
Asparagine	1.777439743	0.086360937	19.69029372
1-Methyl uracil	1.761120907	0.089136716	20.32317114
Quinic acid-like3	1.706176745	0.099045245	22.58231583
Glycine(3TMS)	1.68548275	0.103010222	23.48633071
Glutamic acid	1.675206244	0.10502814	23.94641595
O-Phosphoethanolamine	1.632567987	0.113756987	25.93659312
Glucose_2	1.602463137	0.120276137	27.42295922
Coniferyl aldehyde	1.58131613	0.125037473	28.50854379
Alanine(2TMS)	1.527720569	0.137800873	31.41859912
Fructose_2	1.473112099	0.151874443	34.62737306
Glyceric acid	1.456578493	0.156356168	35.64920621
Fructose_1	1.451747434	0.157685494	35.9522927
Phenylalanine	1.448669622	0.158537089	36.1464564
Threonine(2TMS)	1.43362031	0.162754021	37.10791685
Glucarate_1	1.424934493	0.165228159	37.67202022
Epigallo catechin-like	1.376708139	0.17951188	40.92870871
Maltose_1	1.360278365	0.18459342	42.08729973

(continued)

TABLE 6.5 (CONTINUED)
Result of Significant Test for PCA Loadings

Compound_Name	t Value of PC2 Loading	p Value	Metabolite Number*p Value
Glucono-1,5-lactone_1	1.347783054	0.188532679	42.9854508
Glucose_1	1.337363951	0.191867111	43.74570127
Fumaric acid	1.266888155	0.215632291	49.16416224
Isoleucine	1.136384437	0.265428525	60.51770361
Arabitol	1.128970494	0.2684904	61.21581125
2-Deoxy-D-glucose	1.116153729	0.273843743	62.4363735
Inositol	1.093865243	0.283335562	64.60050816
Gulcono-1,4-lactone	1.089731895	0.285121333	65.00766394
Epigallo catechin	1.089620581	0.285169536	65.01865422
Urea	1.058676542	0.298795169	68.12529862
Galactosamine_1	0.986148648	0.332505633	75.81128427
Pyridoxal_3	0.974404509	0.338199299	77.10944022
Theanine_2	0.971660819	0.339538924	77.41487465
Galactosamine_2	0.97026201	0.340223281	77.57090803
Tryptophan	0.963496914	0.343546198	78.32853313
Maltose_2	0.959141247	0.345697166	78.81895387
Epicatechin	0.929514963	0.360567105	82.20930003
Histidine	0.853383133	0.400688653	91.35701295
Threonine(3TMS)	0.846855991	0.404255714	92.17030285
Tyrosine	0.842766374	0.406500853	92.68219458
Succinic acid	0.810013988	0.424763075	96.84598112
Glucarate_2	0.777847337	0.443182383	101.0455833
Melibiose_1	0.756917547	0.455421752	103.8361595
Leucine	0.741032037	0.464843853	105.9843984
Mannose_2	0.739657974	0.465664184	106.171434
Ornithine	0.734082215	0.469001666	106.9323799
Gallic acid	0.733408154	0.469406082	107.0245868
Quinic acid-like1	0.671487771	0.507412447	115.6900378
Methionine	0.628501386	0.534769279	121.9273957
b-Alanine	0.586407157	0.5623001	128.2044227

TABLE 6.5 (CONTINUED)
Result of Significant Test for PCA Loadings

Compound_Name	t Value of PC2 Loading	p Value	Metabolite Number*p Value
Lysine(4TMS)	0.578203419	0.567748544	129.446668
Sucrose	0.572637689	0.571460088	130.2929001
Putrescine	0.561798532	0.578722997	131.9488434
a-Phenylglycine	0.457113663	0.651116246	148.4545042
Glutamine	0.454098187	0.653258676	148.9429782
2-Aminoethanol	0.399765463	0.692362613	157.8586758
Quinic acid-like2	0.398609341	0.693204648	158.0506598
Pyroglutamic acid	0.389764942	0.699659419	159.5223476
Valine(2TMS)	0.186670292	0.853265105	194.544444
Galactose_2	0.167304858	0.868332656	197.9798455
Aspartic acid	0.146166902	0.884836896	201.7428123
Oxalate	0.111185127	0.912262732	207.995903
Proline	0.053220557	0.957934081	218.4089705
Cyanine	0.029747907	0.976479133	222.6372422
n-Butylamine	0.025148368	0.980114997	223.4662194

Source: Reprinted with permission from Dr. Hiroshi Tsugawa.

same way described in the PCA section: that is, the statistical t value calculation, the p value calculation by "t-dist", and the Bonferroni correction, although we do not explain it here.

  a. Select X variables as shown in the PCA section.
  b. Select "Correlation analysis" radio button.
  c. Click "Done".
  d. Select the file names that you want to examine for relationship. Here, because we want to see the metabolite relationship from high-grade tea to low-grade tea, all samples are selected for the examination.
  e. Click "Apply" button. The correlation analysis can be seen in Figure 6.11.

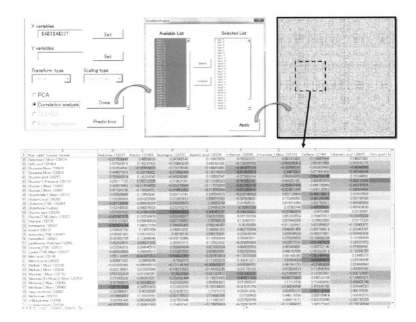

Figure 6.11 Correlation analysis. (Reprinted with permission from Dr. Hiroshi Tsugawa.)

(8) Projection to latent structure regression (PLSR)

   a. Prepare the training data set and test data set.

In this example, for convenience, we utilize No*-1 and No*-3 as training data (i.e., 2012's tea), although No*-2 is utilized as test data (i.e., 2013's tea). The total number of training set and test set is 20 and 10, respectively. From the "Base data" sheet, remove the file name No*-2 and replace the file to another Excel file. As a result, only No*-1 and No*-3 file names remain in the "Base data" sheet. It is important to note that this separation must not be done for the actual analysis and was performed only for the sake of convenience.

   b. Add the *Y*-variable in the last row of "Base data" sheet as described in Figure 6.12.

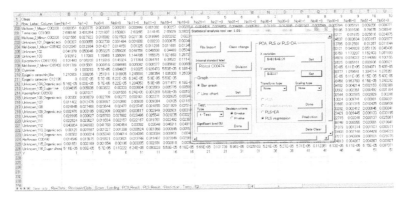

Figure 6.12 How to set *Y*-variables for PLS-based analysis. (Reprinted with permission from Dr. Hiroshi Tsugawa.)

Here, we added the Japanese green tea ranking from the agricultural fair.

    c. Select cell "Y" by clicking the left button of the mouse.

    d. Click the "Set" button.

Note: The program described here supports only 1 *Y* variable.

    e. Select transform type and scaling type.

Here, transform and scaling are "None" and "Auto Scale", respectively.

    f. Click "Done" button.

    g. Optimize the number of latent variables by using the training set and its cross validation result.[12] PRESS (Prediction Residual Error Sum of Squares) or Q2 values are generally utilized to decide the optimal numbers. The Q2 values should be used to decide the number of latent variables although users must not evaluate the model performance by the Q2 values alone. The evaluation of model performance should be done by residual error (RMSEE, RMSEP) as described below.

    h. Add the calculable numbers to the textbox of number of component. And click "Done".

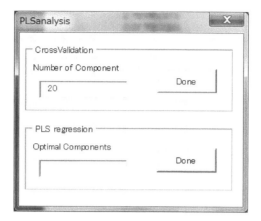

Figure 6.13 Screen shot of cross-validation setting form for PLS analysis. (Reprinted with permission from Dr. Hiroshi Tsugawa.)

Because we utilized the NIPALS algorithm for PLS calculation,[5] we need to set the limit number to stop the calculation. Here we set the textbox to 20 (Figure 6.13).

    i. See the "PLS_Result" sheet.

For calculating the PRESS, Q2, and Q2 cumulative values, this program utilizes the sevenfold cross-validation method. Cross-validation is used to evaluate the model performance without the test set data.[16] The positive values range of Q2 (i.e., the local maximum of Q2 cumulative value or the local minimum of PRESS) should be regarded as the optimal number of latent variables. The Q2 value is defined as the standardized values of PRESS; Q2 cumulative is defined as the infinite product of the Q2 value.

    j. Decide the optimal number of latent variables.

In this example, the local maximum of Q2 cumulative is 4. As a consequence, the optimal value is decided as 4 by using the training set and its cross-validation result (Figure 6.14).

    k. Add the optimal number (4) to the textbox of optimal components and click "Done" button.
    l. Interpret the prediction result, R2 value, and RMSEE values in "PLS_Result" sheet as seen in Figure 6.15.

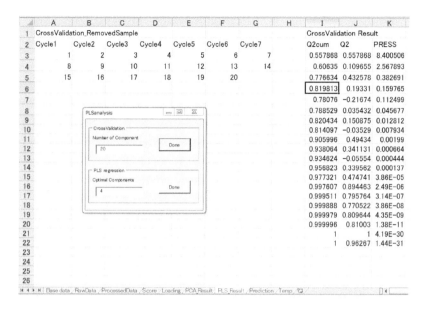

Figure 6.14 Result of cross-validation. (Reprinted with permission from Dr. Hiroshi Tsugawa.)

Here, the graph was made manually and the R2 value was also made using the fitted curve option of Microsoft Excel. In this example, the RMSEE and R2 values are 1.34 and 0.9951, respectively. The RMSEE is the root mean square error of estimation, that is, the so-called standard deviation of the model. As a result of internal validation, the 95%

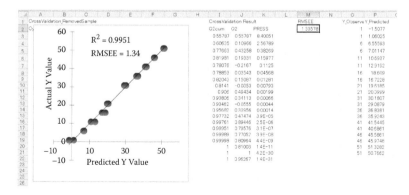

Figure 6.15 Result of internal validation. (Reprinted with permission from Dr. Hiroshi Tsugawa.)

confidence interval of this model was predicted to be the Y value $\pm 2 * 1.34$ (i.e., if the predicted value is 10, the actual value is included between 7.32 and 12.68 with a 95% confidence interval). Researchers should consider whether this performance is practical for the objective quality assessment of Japanese green tea.

m. Extract the significant metabolites correlated to Y variables.

The PLS algorithm always needs three or more latent variables to predict new data sets accurately. It is very difficult to understand simultaneously all loadings used in the model. In order to overcome this problem, the VIP (variable importance in the projection) is frequently used to find the significant metabolites.[17] This is defined as the sum of the multiplied values of the weight loading and the information amount of each latent variable. Conveniently, there is a useful "greater than one rule" that regards more than 1 VIP value as the significant variable involved to Y variables. Our Excel macro generates the VIP values in the "Loading" sheet (Table 6.6). The coefficient value of PLS is also generated in this sheet (Table 6.6).

n. Interpret the VIP values.

TABLE 6.6
VIPs and Coefficients of the Example Data Set

Variable ID	VIP	Coefficient
Raffinose::C00492	2.179926721	0.038144358
Shikimic acid::C00493	1.939973357	0.021628036
Serine_3TMS_Major::C00065	1.844971218	−0.034138181
Arabinose::C00216	1.839247012	0.015233583
Phosphate::C00009	1.825060539	−0.022362018
Maltose_1_Major::C00208	1.819862266	0.043705712
Galactose_1_Major::C00124	1.783563048	0.024309983
Serine_2TMS_Minor::C00065	1.764068442	−0.028254861

(continued)

TABLE 6.6
VIPs and Coefficients of the Example Data Set

Variable ID	VIP	Coefficient
Melibiose_2_Minor::C05402	1.72643739	−0.03142855
threo-3-Hydroxy-L-aspartic acid::C03961	1.684008619	0.005134071
Galactitol::C01697	1.64354983	−0.023744675
Caffeine::C07481	1.616273019	−0.021989818
Xylose_1(or Lyxose_1)_Minor::C00181	1.582938602	−0.038920147
Epicatechin::C09727(8)	1.548734362	−0.035697966
Homoserine::C00263	1.544885806	−0.01892575
Glycine_3TMS_Major::C00037	1.538983217	−0.026331361
Gluconic acid::C00257	1.474052377	0.015010479
2−Isopropylmalic acid::C02504	1.461387493	0.013029741
Xylose_2_Major::C00181	1.444516645	0.013392253
Glucuronate_1_Major::C00191	1.422586011	0.0150987
Lanthionine::PubChem:102950	1.414291376	0.018559678
Malonic acid::C00383	1.406870123	0.022744173
Epigallo catechin_like	1.40522653	−0.020538574
b−Lactose_1_Major::C01970	1.360889227	0.018605041
3,4−dihydroxybenzoate::C00230	1.268628278	0.014945892
Asparagine::C00152	1.229235536	−0.006519889
Adipic acid::C06104	1.202507861	0.020246664
trans−4−Hydroxy−L−proline::C01157	1.202313325	0.021475714
Theanine_2_Same_dif::C01047	1.186904233	−0.015481447
Mannose_1_Major::C00159	1.167506586	0.01263116
Malic acid::C00149	1.150990282	0.004733382
Alanine_2TMS_Major::C00041	1.110959997	−0.016563481
Inositol::C00137	1.096140039	0.029806583
Glyceric acid::C00258	1.084685781	0.005585218
Octadecanoate::C01530	1.055667688	−0.010252398
Quinic acid_like3::C00296	1.0286252	−0.008618171

Source: Reprinted with permission from Dr. Hiroshi Tsugawa.

The extracted metabolites based on the VIP value of the PLS algorithm included all metabolites extracted on the basis of PCA loading. We found highly significant metabolites both in the PCA and PLS. From this, it can be presumed that arabinose (sweet), caffeine (bitter), shikimic acid (sour), and raffinose (sweet) are related to the quality of green tea. Moreover, the coefficient values indicate a negative value for caffeine and positive values for arabinose, shikimic acid, and raffinose (i.e., a high amount of caffeine is included in high-grade green tea and the other metabolites are highly present in low-grade green tea).

o. Validate the model by test set data.

Lastly, validation is needed to determine if the previous model for green tea can be used to predict the quality of the current green tea samples. After optimization of the training set, open the "Prediction" sheet and paste the No*-2 data prepared for test set. Don't forget that the Y values should be added at the end row.

An explanation on how to do the prediction in PLS analysis is shown in Figure 6.16.

p. Click "Prediction" button.
q. Interpret the result shown in Figure 6.17.

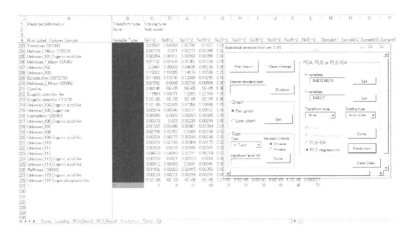

Figure 6.16 How to do the prediction in PLS analysis. (Reprinted with permission from Dr. Hiroshi Tsugawa.)

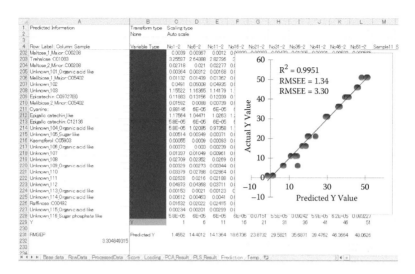

Figure 6.17 Prediction result of PLS analysis. (Reprinted with permission from Dr. Hiroshi Tsugawa.)

RMSEP (root mean square error of prediction) was 3.3. This result shows that the 95% confidence interval is the predicted Y ± 2 * 3.3; that is, if the predicted value is 10, the actual value is included between 3.4 and 16.6 with a 95% confidence interval. As much as the RMSEP value is twice the RMSEE, there may be an overfitting problem in the model. Although the details for model improvement are not explained, the reason for overfitting is possibly due to the low reproducibility of the experiment.

(9) Projection to latent structure discriminant analysis (PLS-DA)

Here, we explain how to use PLS-DA in our Excel macro program.

a. Add the binary vector 0 and 1 for Y variables. In the sample data, we try to distinguish high-grade green tea from the others (i.e., 0 is added to No. 1–No. 6 (high-grade tea) and 1 is added to No. 11–No. 51 (others).

b. Set X and Y variables as seen in Figure 6.18.

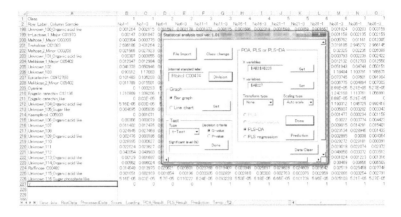

Figure 6.18 How to set the *Y*-variables for PLS-DA. (Reprinted with permission from Dr. Hiroshi Tsugawa.)

c. Select the transform type and scaling type. (In the sample data, transform and scaling types are None and Auto scale, respectively.)

d. Select PLS-DA and click "Done" button.

e. Do cross-validation and decide the optimal number of latent variables. In the sample data, we put 3 as the optimal number (Figure 6.19).

f. Interpret the prediction result and RMSEE value.

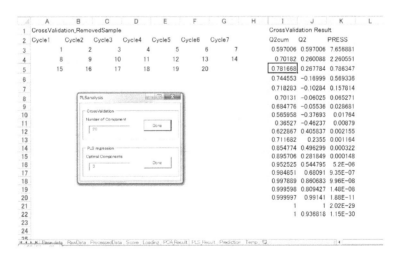

Figure 6.19 How to do PLS-DA. (Reprinted with permission from Dr. Hiroshi Tsugawa.)

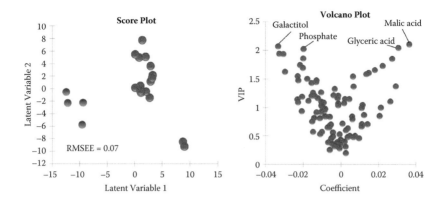

Figure 6.20 Score plot and volcano plot of the PLS-DA result. (Reprinted with permission from Dr. Hiroshi Tsugawa.)

In Figure 6.20, the left figure represents a PLS score plot which is frequently used in a metabolomics study for confirming good separation. On the other hand, a volcano plot shown on the right is a relatively new method in metabolomics study used for finding significant metabolites easily.

    g. Find the significant metabolites using VIP's greater than 1 rule and coefficient values.
    h. Validate the optimal model by using test set.
    i. Interpret the final result.

In the sample data, the result shows that high-grade green tea is perfectly separated from the others by using PLS-DA as can be seen in Figure 6.21.

(10) R-based hierarchical clustering analysis

HCA utilizes just two-dimensional spaces for sample classifications, resulting in easier understanding of biological phenotypes. Here, we introduce how to make a heat map by means of HCA.

    a. Prepare a data matrix in .csv format. Row: Samples Column: Metabolites. Keep the format in Figure 6.22 for easy use.

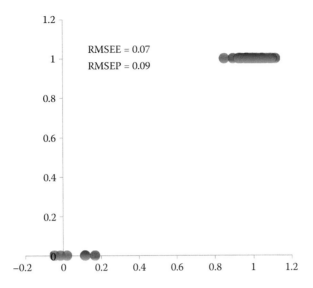

Figure 6.21 Prediction result of PLS-DA. X axis and Y axis show the predicted class and actual class, respectively. (Reprinted with permission from Dr. Hiroshi Tsugawa.)

Figure 6.22 Data file format for HCA. (Reprinted with permission from Dr. Hiroshi Tsugawa.)

b. Download statistical language R (http://cran.r-project.org/bin/windows/base/).[18]
c. Install "gplots"[19] and "amap"[20] libraries. Type the following script in the R editor.

> library(gplots)

> library(amap)

d. Read the .csv format file. Type the following script in the R editor. (But change the directory.)

> data <- read.csv("C:\\Users\\Hiroshi Tsugawa\\
Desktop\\HCA_JapaneseGreenTea.csv", header = T, row.
names = 1")

e. Calculate the distance. "Dist" provides 11 methods for the distance calculations: "euclidean", "maximum", "manhattan", "canberra", "binary", "pearson", "correlation", "spearman", "kendall", "abspearson", and "absorrelation".

> distance_1 <- Dist(data, method = "euclidean")

f. Cluster the distance result. "hclust" provides 7 methods for the clustering: "ward", "single", "complete", "average", "mcquitty", "median", and "centroid".

> cluster_1 <- hclust(distance_1, method = "ward")

g. Create a data matrix to use the "heatmap.2".

> data_matrix <- as.matrix(data)

h. Create a heat map. See the details of each parameter in "heatmap.2" manual.

> heatmap.2(t(data_matrix), scale = "column", dendrogram = "column", Rowv = FALSE, Colv = as.dendrogram(cluster_1), col = greenred(256), main = "Green tea HCA", trace = "none", margin = c(5,15)).

The result of HCA can be seen in Figure 6.23.

    i. If the user wants to use both dendrograms, prepare another cluster as below.

> distance_2 <- Dist(t(data), method = "correlation")

> cluster_2 <- hclust(distance_2, method = "ward")

    j. Create a heat map.

> heatmap.2(t(data_matrix), scale = "column", dendrogram = "both", Rowv = as.dendrogram(cluster_2), Colv = as.dendrogram(cluster_1), col = greenred(256), main = "Green tea HCA", trace = "none", margin = c(5,15))

The HCA result can be seen in Figure 6.24.

6.4 Summary

In this chapter, we introduced the most common statistical analyses in the metabolomics research field using freely available software. In addition to the above analyses, we usually utilize the VANTED[4] and MetaboAnalyst[21] software for pathway map analysis. In the VANTED software, the metabolome data can be easily placed into the pathway map (can be self-constructed or downloaded from a database such as KEGG.[22] In the MetaboAnalyst, we perform the metabolite set enrichment analysis (MSEA) to identify and interpret the metabolite changes in a living organism in a biologically meaningful context.[23]

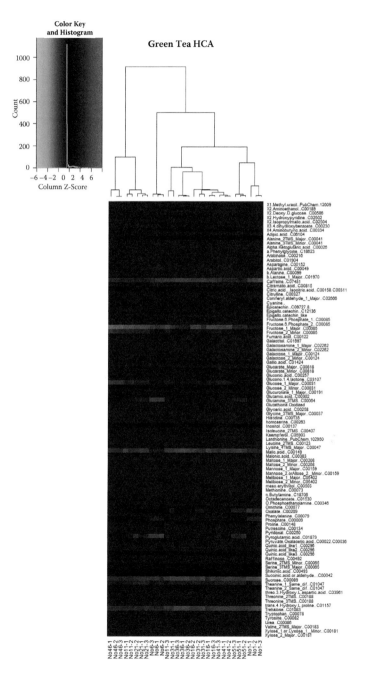

Figure 6.23 HCA result with one axis. (Reprinted with permission from Dr. Hiroshi Tsugawa.)

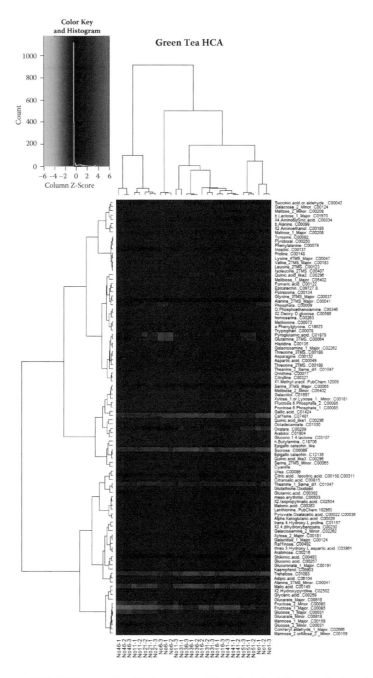

Figure 6.24 HCA result with two axes. (Reprinted with permission from Dr. Hiroshi Tsugawa.)

In order to perform metabolomics study, a researcher must learn analytical chemistry, data processing, and statistical analysis as well as biology. We think that among the topics, the most complicated tasks to learn are data processing and statistical analysis. However, each experimental step in metabolomics includes many systematic errors, therefore, we have to use statistical analysis with correct understanding in order to get novel insights (i.e., find significant metabolites).

In addition, statistical analysis is very important not only for metabolomics study but also for all science-related research fields. There are many kinds of new statistical methods that are now developing all over the world, thus researchers should continue to expand their mathematical knowledge. It is our hope that researchers correctly understand the concept of statistical analysis to avoid making critical misinterpretations of research data.

References

1. Shaffer, J. P., Multiple hypothesis testing: A review, *Annual Review of Psychology*, 1995, 46: 561–584.
2. Yoav, B. and Yosef, H., Controlling the false discovery rate: A practical and powerful approach to multiple testing, *Journal of the Royal Statistical Society B*, 1995, 57: 289–300.
3. Wold, S., Esbensen, K. I. M., and Geladi, P., Principal component analysis, *Chemometrics and Intelligent Laboratory Systems*, 1987, 2: 37–52.
4. Junker, B. H., Klukas, C., and Schreiber, F., VANTED: A system for advanced data analysis and visualization in the context of biological networks, *BMC Bioinformatics*, 2006, 7: 109.
5. Wold, S. and Sjostrom, M., PLS-regression: A basic tool of chemometrics, *Chemometrics and Intelligent Laboratory Systems*, 2001, 58: 109–130.
6. Pérez-Enciso, M. and Tenenhaus, M., Prediction of clinical outcome with microarray data: A partial least squares discriminant analysis (PLS-DA) approach, *Human Genetics*, 2003, 112: 581–592.

7. Semmar, N., Bruguerolle, B., Boullu-Ciocca, S., and Simon, N., Cluster analysis: An alternative method for covariate selection in population pharmacokinetic modeling, *Journal of Pharmacokinetics and Pharmacodynamics*, 2005, 32: 333–358.
8. Daichi, Y., Daisuke, M., Kazunori, S., Katsutoshi, T., and Hiroyuki, W., MALDI–MS-based high-throughput metabolite analysis for intracellular metabolic dynamics, *Analytical Chemistry*, 2010, 82: 4278–4282.
9. Trygg, J. and Wold, S., Orthogonal projections to latent structures, *Journal of Chemometrics*, 2002, 16: 119–128.
10. Bylesjö, M., Rantalainen, M., Cloarec, O., Nicholson, J. K., Holmes, E., and Trygg, J., OPLS discriminant analysis: Combining the strengths of PLS-DA and SIMCA classification, *Journal of Chemometrics*, 2007, 341–351.
11. Wiklund, S., Johansson, E., Sjöström, L., Mellerowicz, E. J., Edlund, U., Shockcor, J. P., Gottfries, J., Moritz, T., and Trygg, J., Visualization of GC/TOF-MS-based metabolomics data for identification of biochemically interesting compounds using OPLS class models, *Analytical Chemistry*, 2008, 80: 115–122.
12. Dunn, W. B., Broadhurst, D., Begley, P., Zelena, E., Francis-McIntyre, S., Anderson, N., Brown, M., Knowles, J. D., Halsall, A., Haselden, J. N., Nicholls, A. W., Wilson, I. D., Kell, D. B., and Goodacre, R., Procedures for large-scale metabolic profiling of serum and plasma using gas chromatography and liquid chromatography coupled to mass spectrometry, *Nature Protocols,* 2011, 6: 1060–1083.
13. Jonsson, P., Gullberg, J., Nordström, A., Kusano, M., Kowalczyk, M., Sjöström, M., and Moritz, T., A strategy for identifying differences in large series of metabolomic samples analyzed by GC/MS, *Analytical Chemistry*, 2004, 76: 1738–1745.
14. van den Berg, R., Hoefsloot, H. C. J., Westerhuis, J., Smilde, A. K., and van der Werf, M. J., Centering, scaling, and transformations: Improving the biological information content of metabolomics data, *BMC Genomics,* 2006, 7: 142.
15. Hiroyuki, Y., Tamaki, F., Hajime, S., Gen, I., and Yoshiaki, O., Statistical hypothesis test of factor loading in principal component analysis and its application to metabolite set enrichment analysis, 2012 International Conference of the Metabolomics Society.
16. Svante, W., Pattern recognition by means of disjoint principal components models, *Pattern Recognition*, 1976, 8: 127–139.

17. Gosselin, R., Rodrigue, D., and Duchesne, C., A bootstrap-VIP approach for selecting wavelength intervals in spectral imaging applications, *Chemometrics and Intelligent Laboratory Systems*, 2010, 100: 12–21.

18. R Development Core Team ISBN 3-900051-07-0, URL http://www. R-project.org, 2005.

19. Gregory, R. W., Ben, B., Lodewijk, B., Robert, G., Wolfgang, H. A. L., Marc, S., and Bill, V., R package version 2.12.1. http://CRAN.R-project.org/package=amap, 2013.

20. Antoine, L., R package version 0.8-7. http://CRAN.R-project.org/package=amap, 2011.

21. Xia, J. and Wishart, D. S., Web-based inference of biological patterns, functions and pathways from metabolomic data using MetaboAnalyst, *Nature Protocols*, 2011, 6: 743–760.

22. Ogata, H., Goto, S., Sato, K., Fujibuchi, W., Bono, H., and Kanehisa, M., KEGG: Kyoto Encyclopedia of Genes and Genomes, *Nucleic Acids Research,* 1999, 27: 29–34.

23. Xia, J. and Wishart, D. S., MSEA: A web-based tool to identify biologically meaningful patterns in quantitative metabolomic data, *Nucleic Acids Research*, 2010, 38: W71–7.

7
Case Studies

Chapter 7

Case Studies

Walter A. Laviña, Yusuke Fujieda, Udi Jumhawan,
Sastia Prama Putri, and Eiichiro Fukusaki

Chapter Outline

7.1 Metabolomics as a Tool for Prediction of Phenotypes

Case Study: Metabolomics-based systematic prediction of yeast lifespan and its application for semi-rational screening of ageing-related mutants[1]

7.1.1 Introduction

Functional genomics, in its classical definition, is the high-throughput technology-based clarification of gene functions and interactions using "dynamic aspects" such as gene transcription, translation, and protein–protein interaction.[2–4] More recently, metabolomics, the systematic and exhaustive profiling of metabolites involving multiple disciplines, is becoming a very important tool in functional genomics for elucidating genetic and cellular functions.[5] Most scientists believe that the integrative experimental approach, combining different "omics" fields (genomics, transcriptomics, proteomics, metabolomics), is the best approach for comprehensive understanding of gene functions.[6] However, this same approach is sometimes a limiting factor for the full potential of metabolomics and inasmuch as the metabolome is considered as a result of the execution of the genetic information, metabolomics is deemed an important tool for characterizing phenotypes. In fact, this premise has been demonstrated in discriminating silent mutants through the use of metabolic fingerprinting and footprinting.[7,8] These results show that metabolomics can distinguish very similar phenotypes with very high resolution. Metabolic profiling can also discriminate between normal and abnormal cells for finding disease-related biomarkers.[9–11] As machines and equipment for acquiring metabolic information become more advanced and complex, the metabolic data obtained become more comprehensive, thus leading to more accurate explanation of biological functions. Recently, Yoshida and colleagues[1] have developed an entirely different approach in utilizing metabolomics for characterization of phenotypes. In this study, quantitative changes in replicative lifespan were assessed in yeast strains using metabolic profiles acquired from GC/MS and CE/MS data (Figure 7.1). In this chapter, the use of metabolomics for the discrimination of quantitative changes and systematic prediction of a quantitative phenotype (longevity in *Saccharomyces cerevisiae*) is described.

7.1.2 Results

7.1.2.1 Identification and Quantification of Yeast Metabolites

Replicative life span (RLS), defined as the number of times an individual cell divides before undergoing senescence, was used in this study for demonstrating the capability of metabolic profiling to differentiate changes in a quantitative

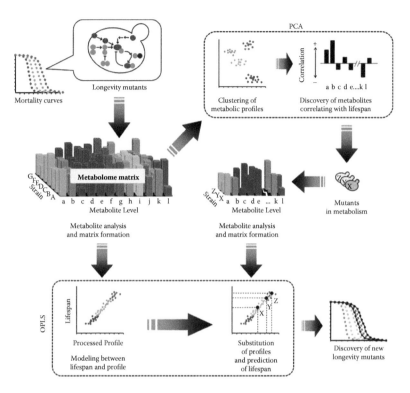

Figure 7.1 Model of a novel method for the identification of lifespan extension mutants using metabolic profiling. The metabolomes of strains with distinct lifespans were profiled and multivariate analysis of the metabolome data showed that lifespan extension correlated with metabolic changes, thus providing the basis for a method for lifespan prediction. This approach was able to strongly complement the discovery of new longevity mutants suggesting that metabolomics can successfully analyze measurable phenotypes. (From Yoshida et al., *Aging Cell*, 2010, 9: 616–625. With permission from John Wiley and Sons.)

phenotype (in this case, longevity).[12] Specifically, six yeast strains with different replicative lifespans were used in the study, namely wild type (BY4742), the short-lived mutant *pde2Δ*, and the long-lived mutants *hxk2Δ*, *idh2Δ*, *gpa2Δ*, and *tor1Δ*.[13,14] For the construction of metabolic profiles, a total of 87 low-molecular–weight hydrophilic compounds in the budding yeast *Saccharomyces cerevisiae* were identified by gas chromatography/mass spectrometry (GC/MS) and capillary electrophoresis-mass spectrometry (CE/MS). In the wild-type strain (BY4742 background) grown in YPD medium, these

TABLE 7.1
**Low-Molecular–Weight Compounds Identified in *S. cerevisiae*
by GC/MS and CE/MS for Construction of Metabolic Profiles**

Category	Metabolites Detected by Mass Spectrometries	Detection System
Amino acid	4-Aminobutyrate	GC/MS
Amino acid	5-Aminolevulinate	GC/MS
Amino acid	Alanine	GC/MS
Amino acid	Allothreonine	GC/MS
Amino acid	Asparagine	GC/MS
Amino acid	Aspartate	GC/MS
Amino acid	Citrulline	GC/MS
Amino acid	Cysteine	GC/MS
Amino acid	Glutamate	GC/MS
Amino acid	Glutamine	GC/MS
Amino acid	Glycine	GC/MS
Amino acid	Histidine	GC/MS
Amino acid	Homoserine	GC/MS
Amino acid	Isoleucine	GC/MS
Amino acid	Lysine	GC/MS
Amino acid	Methionine	GC/MS
Amino acid	Norleucine	GC/MS
Amino acid	Ornithine	GC/MS
Amino acid	Phenylalanine	GC/MS
Amino acid	Proline	GC/MS
Amino acid	Pyroglutamate	GC/MS
Amino acid	Serine	GC/MS
Amino acid	Threonine	GC/MS
Amino acid	Tryptophane	GC/MS
Amino acid	Tyrosine	GC/MS
Amino acid	Valine	GC/MS
Nucleotide/Nucleotide derivative	Adenine	GC/MS
Nucleotide/Nucleotide derivative	Adenosine	GC/MS
Nucleotide/Nucleotide derivative	ADP	CE/MS

TABLE 7.1 (CONTINUED)
Low-Molecular–Weight Compounds Identified in *S. cerevisiae*
by GC/MS and CE/MS for Construction of Metabolic Profiles

Category	Metabolites Detected by Mass Spectrometries	Detection System
Nucleotide/Nucleotide derivative	AMP	CE/MS
Nucleotide/Nucleotide derivative	cAMP	CE/MS
Nucleotide/Nucleotide derivative	CMP	CE/MS
Nucleotide/Nucleotide derivative	GMP	CE/MS
Nucleotide/Nucleotide derivative	Hypoxanthine	GC/MS
Nucleotide/Nucleotide derivative	Inosine	GC/MS
Nucleotide/Nucleotide derivative	TMP	CE/MS
Nucleotide/Nucleotide derivative	UDP	CE/MS
Nucleotide/Nucleotide derivative	UMP	CE/MS
Nucleotide/Nucleotide derivative	Uracil	GC/MS
Organic acid	2-Oxoglutarate	GC/MS
Organic acid	2-Oxoglutarate	CE/MS
Organic acid	Citrate	GC/MS
Organic acid	Citrate	CE/MS
Organic acid	Fumarate	CE/MS
Organic acid	Glycerate	GC/MS
Organic acid	Glycolate	GC/MS
Organic acid	Glyoxylate	GC/MS
Organic acid	Lactate	GC/MS
Organic acid	Malate	GC/MS
Organic acid	Malate	CE/MS
Organic acid	Oxalate	GC/MS
Organic acid	Pantothenate	CE/MS
Organic acid	Pyruvate	GC/MS
Organic acid	Shikimate	GC/MS
Organic acid	Succinate	GC/MS
Organic acid	Succinate	CE/MS
Other	Laurate	GC/MS
Other	2-Hydroxypyridine	GC/MS

(*continued*)

TABLE 7.1 (CONTINUED)
Low-Molecular–Weight Compounds Identified in *S. cerevisiae*
by GC/MS and CE/MS for Construction of Metabolic Profiles

Category	Metabolites Detected by Mass Spectrometries	Detection System
Other	Acetyl-CoA	CE/MS
Other	alpha-Glycero-P	CE/MS
Other	beta-Glycero-P	CE/MS
Other	Ethanolamine	GC/MS
Other	FAD	CE/MS
Other	FMN	CE/MS
Other	Glycerol	GC/MS
Other	NAD	CE/MS
Other	n-Butylamine	GC/MS
Other	Orotate	GC/MS
Other	Phosphate	GC/MS
Other	Putrescine	GC/MS
Sugar/Sugar alcohol	3-Phosphoglycerate	CE/MS
Sugar/Sugar alcohol	ADP-ribose	CE/MS
Sugar/Sugar alcohol	Dihydroxyacetone-P	CE/MS
Sugar/Sugar alcohol	Erythritol	GC/MS
Sugar/Sugar alcohol	Fructose-1,6-bisP	CE/MS
Sugar/Sugar alcohol	Glucose	GC/MS
Sugar/Sugar alcohol	Glucose-1-P	CE/MS
Sugar/Sugar alcohol	Hexose-6-P	CE/MS
Sugar/Sugar alcohol	Inositol	GC/MS
Sugar/Sugar alcohol	Mannose	GC/MS
Sugar/Sugar alcohol	Phosphoenol pyruvate	CE/MS
Sugar/Sugar alcohol	Ribose	GC/MS
Sugar/Sugar alcohol	Ribose-5-P	CE/MS
Sugar/Sugar alcohol	Ribulose-5-P	CE/MS
Sugar/Sugar alcohol	Sedoheptulose-7-P	CE/MS
Sugar/Sugar alcohol	Trehalose-6-P	CE/MS
Sugar/Sugar alcohol	UDP-glucose	CE/MS

Source: Yoshida et al., *Aging Cell*, 2010, 9: 616–625. With permission from John Wiley and Sons.

included 55 compounds such as amino acids, organic acids, sugars, sugar alcohols, and 32 compounds such as phosphory-lated sugars, nucleotides, and coenzyme related compounds identified using GC/MS and CE/MS, respectively (Table 7.1).

To visualize the multivariate data of the profiles expressed as relative levels of the 87 metabolites, principal component analy-sis (PCA) was used. The variance along PC1 (47% of the total variance) showed separation of the wild-type and mutant strains grown in 2% glucose from the wild type grown in 0.5% glucose under calorie restriction (CR), indicating that CR affects meta-bolic profiles of the yeast strains. On the other hand, the variance along PC2 (13% of the total variance) was shown to be related to the lifespan of wild-type and mutant strains. Specifically, higher PC2 scores were a characteristic of long-lived mutants and the calorie-restricted wild type as compared to wild type grown in 2% glucose. In addition, the short-lived mutant *pde2Δ* showed a lower PC2 score compared to all others (Figure 7.2A).

7.1.2.2 Metabolites Associated with Lifespan Extension

Metabolites that contributed to the separation of the strains along PC2 (in the PCA) based on lifespan were correlated with lifespan extension. The relevant loading plots of PC2 corre-spond to the relative degree of correlation between the level of each metabolite and lifespan. Specifically, lifespan expan-sion was found to correlate positively with high levels of amino acids derived from aspartate and glutamate (methionine, pro-line, threonine, isoleucine, histidine, glutamine, and aspartate) whereas high levels of nucleotides and their derivatives (inosine, hypoxanthine, cAMP, TMP, GMP, and ribose) were negatively correlated with lifespan extension. This indicated that amino acids and nucleotide-related compounds exert positive and neg-ative effects on lifespan extension, respectively (Figure 7.2B).

7.1.2.3 Construction of the Lifespan Prediction Model

Previously, both qualitative and quantitative analyses of lifespan were described and based on these, a mathematical model that predicts the yeast lifespan based on metabolic pro-files was constructed. The correlation between the replicative

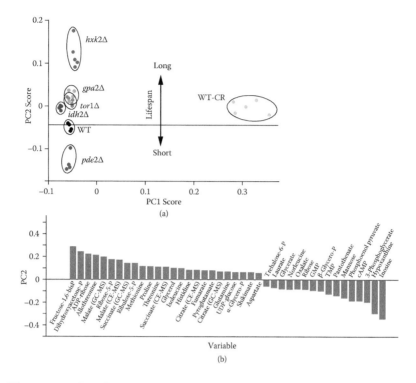

Figure 7.2 Correlations between metabolic profiles and lifespan variation. (a) Yeast lifespan discrimination based on metabolomics. A principal
components analysis score plot showing metabolites in yeast cells that are
affected CR or longevity mutations. Each point represents individual batch
color-coded according to lifespan. Percent effect on the mean lifespan relative to the wild-type strain: wild type under CR (WT-CR), 29; hxk2Δ, 34;
idh2Δ, 32; gpa2Δ, 31; tor1Δ, 20; pde2Δ, −29. (b) 41 metabolites most important to lifespan discrimination. A loadings plot corresponding to the scores
plot from the PCA shows the variables making the largest contributions to
PC2 and the direction of each contribution. The upper and lower loading
values of the metabolites were above 0.05 and below −0.05, respectively.
The mass spectrometry method by which each metabolite was detected is
indicated in parentheses when both methods detected the same metabolite.
ADP-ribose, adenosine diphosphate ribose; UDP-glucose, uridine diphosphate glucose; GMP, guanosine monophosphate; TMP, tymidine monophos-
phate; cAMP, cyclic adenosine monophosphate. "P" indicates phosphate.
(From Yoshida et al., *Aging Cell*, 2010, 9: 616–625. With permission from
John Wiley and Sons.)

lifespans and metabolic profiles of the wild-type and mutant strains in PCA was analyzed using an orthogonal projection to latent structure (OPLS) algorithm and cross-validation was done by comparison of the OPLS-predicted lifespan value of a test strain with the actual lifespan value. Using two predictive components for regression modeling, results showed a good correlation between lifespan and the fingerprint obtained from metabolic profiling. The model included 94% of the variation in lifespan ($R2 = 0.94$) and predicted 91% of the variation in lifespan ($Q2 = 0.91$; Figures 7.3A,B). Furthermore, the metabolites responsible for lifespan discrimination in PCA such as nucleotide-related compounds (TMP, GMP, inosine, and hypoxanthine) and amino acids (methionine, proline, histidine, and glutamine) were also considered important for the construction of the prediction model. Owing to the fact that the OPLS model was robust after four rounds of cross-validation (Figure 7.3C), it was concluded that the prediction model can estimate the replicative lifespan of an uncharacterized mutant strain without direct measurement.

7.1.2.4 Prediction of Lifespans for the Discovery of Novel Longevity Mutants

From the results of the PCA, seven genes (two involved in amino acid metabolism and five in nucleotide metabolism) that encode for gene products in the metabolic pathways of the compounds correlated with lifespan were selected for further studies (Table 7.2). After analyzing the metabolites of the seven lifespans' mutant strains using GC/MS and CE/MS, their respective lifespans were predicted using the OPLS prediction model. Results of the model roughly approximated a +14% to +28% increase in lifespan when individual genes are deleted which was confirmed by direct measurement of the lifespan of the mutants (Table 7.3). In particular, *uga3Δ* (GABA utilization activator), *urh1Δ* (uridine nucleosidase), and *fzf1Δ* (sulphite metabolism activator) mutants exhibited significant extension of their lifespan

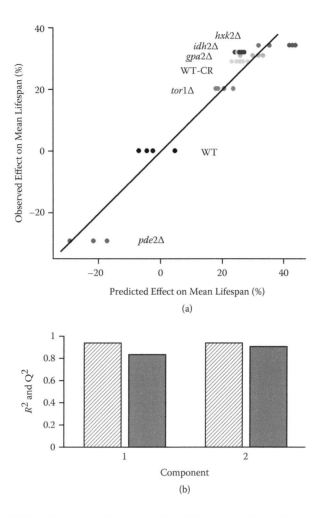

Figure 7.3 Prediction of the replicative lifespans of long-lived mutants. (a) Observed vs. Predicted lifespan values for a two-component orthogonal projection to latent structure model in which all predictions relate to mode-building data. The data were fit with a linear regression line with a slope of 1.0. Each value is represented as the percent effect on the mean lifespan relative to the wild-type strain. Each point represents an individual batch, with color-coding according to lifespan. (b) Performance of the predictive model. R2 (hatched) is the proportion of the sums of squares explained by the model and describes the degree of agreement between the derived model and the data. Q2 (shaded) is the cross-validated R2 and describes the predictive ability of the derived model. In general, Q2 > 0.9 is regarded as excellent, and the difference between R2 and Q2 must not exceed 0.3. (From Yoshida et al., *Aging Cell*, 2010, 9: 616–625. With permission from John Wiley and Sons.) (*continued*)

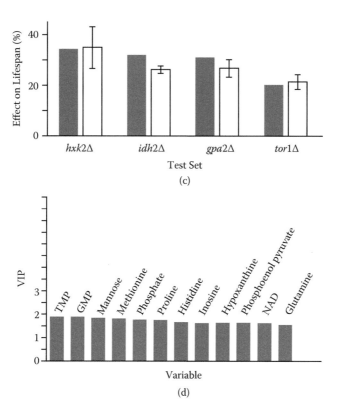

Figure 7.3 (continued) Prediction of the replicative lifespans of long-lived mutants. (c) The result of a four-round cross-validation in which each mutant represents the test data not used in model building (black columns, observed effect on mean lifespans; white columns, predicted effect on mean lifespans +/–SD, n = 4 or 5). (d) The 12 metabolites most important to the predictive model. VIP, variable influence on projection. (From Yoshida et al., *Aging Cell*, 2010, 9: 616–625. With permission from John Wiley and Sons.)

whereas the remaining four mutants did not show considerable effect in longevity (Figures 7.4A,B). Seemingly, the maximum threshold for prediction ability of the OPLS model in this study was 25%; as the predicted lifespan extension rates above and below 25% correspond to increased and wild-type comparable lifespans, respectively. The results of this study demonstrate a novel approach for effectively discovering longevity mutants based on metabolic profiling, PCA, and OPLS analyses.

TABLE 7.2
Genes Presumed to Be Related to Lifespan

Gene	Protein Function	Related Pathway	Metabolites in Pathway Correlated with Lifespan Extension[a]
URH1	Uridine nucleosidase	Salvage of pyrimidine nucleotide	TMP, ribose-5-P, ribose
UGA3	Transcriptional activator	Glutamate metabolism	Proline, glutamate, histidine
FZF1	Transcription factor involved in sulphite metabolism	Synthesis of cysteine and methionine	Methionine, aspartate
TAD1	Adenosine deaminase	Salvage of purine nucleotide	Inosine, hypoxanthine, GMP
FCY1	Cytosine deaminase	Salvage of pyrimidine nucleotide	TMP, ribose-5-P, ribose
IMD1	Inosine monophosphate dehydrogenase	Salvage of purine nucleotide	Inosine, hypoxanthine, GMP
SML1	Ribonucleotide reductase inhibitor	Synthesis of deoxyribonucleotide	Inosine, hypoxanthine, GMP, TMP, ribose-5-P, ribose

Source: Yoshida et al., 2010, *Aging Cell*, 9: 616–625. With permission from John Wiley and Sons.

[a] For metabolites listed in this column, refer to Figure 7.2B.

7.1.3 Discussions

7.1.3.1 Quantitative Association Between Metabolome and Lifespan

From visual inspection of the PCA score plot along PC2, the clusters of long-lived mutants and wild type grown under CR showed distinct partition from the wild-type strain; supporting the previous assertion that *hxk2Δ*, *gpa2Δ*, and *tor1Δ* mutants have phenotypes similar to CR strains (Figure 7.2A).[14,15] In addition, strains with longer lifespans showed higher scores in

TABLE 7.3
Lifespan Prediction and Determination

Gene	Lifespan Extension Rate When Gene Is Deleted[a]		
	Predicted Rate	**Measured Rate**	**P-Value**
URH1	28 ± 5	24*	1.6×10^{-3}
UGA3	28 ± 4	75*	$< 0.1 \times 10^{-6}$
FZF1	25 ± 2	66*	0.1×10^{-6}
TAD1	22 ± 6	14	0.014
FCY1	22 ± 2	−4	0.92
IMD1	21 ± 1	−1	0.66
SML1	14 ± 6	3	0.51

Source: Yoshida et al., 2010, *Aging Cell*, 9: 616–625. With permission from John Wiley and Sons.

[a] Each value is represented as percent effect on the mean lifespan relative to the wild-type strain. Predicted rate is shown with mean value ± SD ($n = 5$). Asterisks indicate statistically significant difference between wild type and mutant strains ($n = 48$, $P < 0.01$; Wilcoxon rank-sum test).

the PC2 coordinate, thus these results suggest that the metabolic profile of each strain is related to its lifespan.

7.1.3.2 PCA Revealed Metabolites Associated with Lifespan

PCA revealed that the levels of certain metabolites were related to lifespan extension. In particular, an increase in amino acids and intermediates of the tricarboxylic acid cycle (malate, succinate, citrate, and fumarate) correlated positively with lifespan extension. Although amino acids were previously reported to prevent lifespan extension when present in the growth medium, intracellular amino acids may not exhibit the same effect.[14,16] In the case of the TCA intermediates, the findings of this study support the earlier observation that lifespan-extending CR and its mimetics enhance respiration.[17,18]

Although the use of metabolic fingerprints and footprints to classify yeast mutants has been reported before, individual compounds were not identified in these studies.[7,8] In contrast, this study was able to identify and quantify individual metabolites associated with replicative lifespan and using this

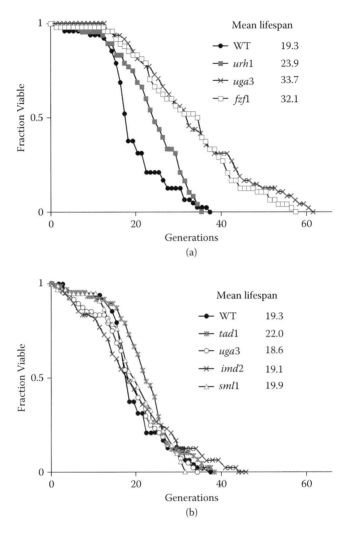

Figure 7.4 Replicative lifespans of the strains as predicted by the model. (a) Among the mutants with lifespans that were predicted using orthogonal projection to latent structure (OPLS), uga3Δ, urh1Δ, and fzf1Δ had significantly increased lifespans. (b) Lifespans of tad1Δ, fcy1Δ, imd2Δ, and sml1Δ were not increased significantly. (From Yoshida et al., *Aging Cell*, 2010, 9: 616–625. With permission from John Wiley and Sons.)

information, longevity mutants were characterized by focusing on the metabolites with high correlation with lifespan. Based on this, it was deduced that deletion of genes related to these metabolites could alter lifespan.

7.1.3.3 Novel Application of the OPLS-Based Prediction Model

In the PCA, mutant strains clustered according to lifespan, thus it was inferred that metabolic profiles could be used for prediction of mutants with uncharacterized lifespan. This hypothesis was confirmed by OPLS analysis, which plotted the quantitative lifespan measurements against the metabolic profiles. A validated model predicting yeast lifespan was constructed based on this correlation that accurately predicted longevity of four longevity mutants during cross-validation. This model was able to identify three novel longevity mutants out of seven deletion mutants tested (42.8%) as compared to 2.3% discovered previously using random screening of single gene deletion strains.[14] Although the reason for inaccurate prediction of four lifespan mutants is unclear, the accuracy of the prediction model is likely to increase with a corresponding increase in the volume of metabolic data used for model construction.

A simplified, high-throughput screening method for analysis of lifespan without the use of micromanipulation has been reported before, which showed that the viability of a population displays features of a survival curve such as changes in hazard rate with age.[19] This method, however, takes several days to complete because measurement of ageing of a cohort requires time to complete. In contrast, OPLS takes less time to predict lifespan and can be accomplished within a day if cell samples are prepared in advance.

Using a novel metabolomics approach, this study was able to represent quantitatively a measurable phenotype with a metabolic profile from which metabolic pathways affecting phenotypic variation in lifespan were deduced. Furthermore, a method for longevity prediction was developed for screening longevity mutants. The semirational screening method presented in this study aims to maximize the comprehensive analysis of lifespan because the methods used in previous studies of replicative lifespan are time-consuming and laborious. This was done using two processes: (1) PCA-based identification of metabolites with levels that are highly correlated with lifespan, followed by the selection of genes relevant to their metabolism; and (2) OPLS-based prediction of the effect of longevity of deletions of these genes.[1]

In this study, relationships among metabolome, phenotype, and longevity allowed prediction of relevant character traits. This quantitative approach provides a strategy for identifying and efficiently screening candidate mutation affecting otherwise genetically intractable phenotypes.[1]

7.2 Metabolomics as a Tool for Discrimination and Authentication

Case study: Selection of Discriminant Markers for Authentication of Asian Palm Civet Coffee (Kopi Luwak): A Metabolomics Approach.[20]

7.2.1 Introduction

Coffee is one of the most popular beverages in the world and among a wide variety of types, Kopi Luwak is considered to be the world's most expensive coffee, with a price tag of $300–400 USD per kg.[21] Kopi Luwak, whose name is derived from Indonesian words for coffee and civet, respectively, is made from coffee berries that have been eaten by the Asian palm civet (*Paradoxurus hermaphroditus*), a small mammal native to Southern and Northern Asia.[22] The civet climbs coffee trees and instinctively selects coffee cherries. During digestion, the coffee pericarp is completely digested and the beans are excreted. The intact beans are then collected, cleaned, wet-fermented, sun-dried, and further processed by roasting. Kopi Luwak's high selling price is mainly attributed to its exotic and unexpected production process.[23]

Despite its profitable prospects, there is no reliable standardized method for determining the authenticity of Kopi
Luwak. Moreover, there is limited scientific information on
this exotic coffee. Recently, coffee adulterated to resemble Kopi
Luwak was reported in the coffee market.[24] This poses serious
concern among consumers over the authenticity and quality of
the products currently available in the market. Discrimination
between Kopi Luwak and regular coffee has been achieved
using electronic nose data.[23] However, the selection of a discriminant marker for authentication was not addressed. The
method currently employed by Kopi Luwak producers is sensory analysis that includes visual and olfactory testing, both
of which are inadequate. For example, visual examination is
only possible for green coffee beans prior to roasting, and very
few trained experts can perform the highly subjective sensory
analysis to discriminate Kopi Luwak.

Information flow in metabolic pathways is highly dynamic
and represents the current biological states of individual
cells. Hence, the metabolome has been considered as the best
descriptor of physiological phenomena.[25] Metabolomics techniques can be powerful tools to elucidate variations in phenotypes imposed by perturbations such as gene modification,
environmental factors, or physical stress. The "black box" process during animal digestion can be translated as physical and
enzymatic consequences to the coffee bean, which presents a
smoother surface and color changes after digestion.[23] Thus, a
metabolomics technique was chosen to screen and select discriminant markers for the authenticity assessment of Kopi
Luwak. Metabolomics techniques have been effectively applied
to distinguish the phytochemical compositions of agricultural
products of different origins,[26] varieties,[26,27] and cultivars[28] for
quality control and breeding.

In this study, gas chromatography coupled with quadrupole
mass spectrometry (GC/Q/MS)-based multimarker profiling
was employed to identify discriminant markers for the differentiation of Kopi Luwak and regular coffees. A combination of
gas chromatography and mass spectrometry (GC/MS) provides
high sensitivity, reproducibility, and the quantitation of a large
number of metabolites with a single-step extraction.[29,30] Sample
classification by means of chemometrics was performed using

principal component analysis. Subsequently, orthogonal projection to latent structures combined with discriminant analysis (OPLS-DA),[31] and significance analysis of microarrays/metabolites (SAM)[32] were used to isolate statistically significant compounds as discriminant marker candidates. The applicability of these candidates as discriminant markers was verified to differentiate various commercial coffee products.

7.2.2 Results and Discussion

7.2.2.1 GC/MS-Based Metabolite Profiling of Kopi Luwak

GC/Q/MS analysis was performed on aqueous extracts of coffee beans to investigate the differences in their metabolite profiles and select discriminant markers for robust authentication. In addition, this research focused on increasing the scientific information about Kopi Luwak. A quadrupole mass spectrometer was selected because of its availability as the most widely used mass analyzer. However, a conventional Q/MS can be operated only at a slow scan rate.[33] With processor improvements and high-speed data processing, newly developed GC/Q/MS instruments provide increased sensitivity at high scan speeds of up to 10.000 u/s.[34]

Because of their broad cultivation areas and commercial profitability, *Coffea arabica* and *Coffea canephora*, which represent 65% and 35% of the total annual coffee trade, respectively, were utilized for metabolomics analysis.[35] A total of 182 peaks from 21 coffee beans was extracted using MetAlign. Moreover, 26 compounds were tentatively identified by comparison with our in-house library (by retention index) and the NIST library (by retention time); six of these were identified by coinjection with an authentic standard. Tentatively identified components consisted of organic acids, sugars, amino acids, and other compounds. Previously reported coffee bean constituents, including chlorogenic, quinic, succinic, citric, and malic acids; caffeine, one of the compounds supplying the bitter taste in coffee; and sucrose, the most abundant simple carbohydrate, were identified.[36-40]

In recent research, unsupervised analysis, PCA, has been employed for data exploration and visualization of information based on sample variance.[41,42] A PCA score plot derived from the

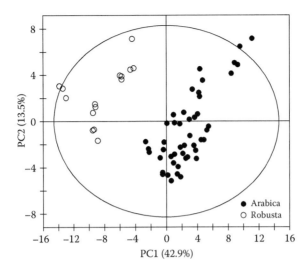

Figure 7.5 PCA score plot of experimental coffee set. Coffea arabica and Coffea canephora (Robusta) were clearly separated. Genetic diversity strongly influenced data separation. (From Jumhawan et al., *Journal of Agricultural and Food Chemistry,* 2013, 61: 7994–8001. With permission from ACS.)

21 coffee beans differentiated two data groups based on their species, Arabica and Robusta (Figure 7.5), and resulted in a goodness-of-fit parameter (R2) equal to 0.844. Caffeine and quinic acid were significant for the Robusta coffee data set, whereas the Arabica data set was mainly supported by various organic acids such as malic, chlorogenic, citric, and succinic acids. The data differentiation was explained by 42.9% variance along PC1. The results indicated that genetic diversity more strongly influenced the data separation than animal perturbation.

Because of the large variance among coffee species, sample differentiation based on the type of coffee, whether it is Kopi Luwak or regular coffee, could not be observed. Additional analyses were carried out independently for each coffee species originating from the same cultivation area. The PCA score plot revealed data separation based on the type of coffee, in which Kopi Luwak and regular coffee could be clearly separated (Figure 7.6). For the Arabica coffee data set, the separation was explained by 45.5% and 27.7% variances in PC1 and PC2, respectively. In PC2, Kopi Luwak was closely clustered

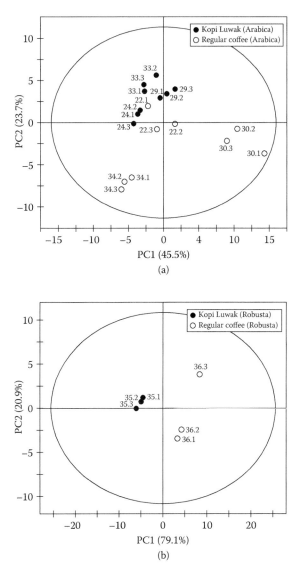

Figure 7.6 PCA score plot of Kopi Luwak and regular coffee from same cultivation area, Arabica (A) and Robusta (B). (From Jumhawan et al., *Journal of Agricultural and Food Chemistry,* 2013, 61: 7994–8001. With permission from ACS.)

in the same region, whereas regular coffees tended to separate based on their cultivation areas. In the loading plot, malic and glycolic acids contributed highly to the Kopi Luwak data (data not shown). Thereby, coffee beans may possess similar profiles after animal digestion. Differences in cultivation areas were considered to have the least significance for data separation. In Robusta coffee, a clear separation between Kopi Luwak and regular coffee was observed, which was explained by 79.1% variance of PC1. Significant compounds for separation, including inositol and pyroglutamic acid for Kopi Luwak and quinic acid for regular coffee, were observed.

7.2.2.2 Discriminant Analysis to Select Candidates for Discriminant Markers

An overview of all data samples was provided by the unsupervised analysis, PCA. However, detailed information regarding compounds contributing to the data differentiation between Kopi Luwak and regular coffee remained unclear. Therefore, coffee bean data sets were subjected to supervised discriminant analysis. For analyses having two or more classes, OPLS-DA is the most suitable platform for isolating and selecting differentiation markers. Compounds with reliable high contributions to the model may possess potentially biochemically interesting characteristics; thus they can be selected as biomarker candidates.[31] All OPLS-DA models exhibited R2 and Q2 values greater than 0.8, which would be categorized as excellent.[43] In addition, all models were in the range of validity after permutation tests using 200 variables (data not shown). The model was considered valid after permutation for those that met the following criteria: R2Y-intercepts and Q2-intercepts that did not exceed 0.3–0.4 and 0.05, respectively.[44]

Potential candidates for discriminant markers can be selected via S-plots by setting the cutoff for covariance, $p[1]$, and the correlation value, $p[corr]$, to $> |0.2|$. S-plots of the coffee data sets are shown in Figures 7.7B and 7.7D. In addition to cutoff values, candidates for discriminant markers were selected by variable importance in projection (VIP) values. Large VIP values (>1) are more relevant for model construction.

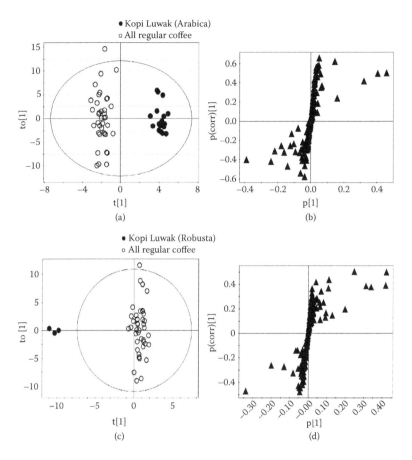

Figure 7.7 OPLS-DA score plots (a, c) and loadings of S-plots (b, d) derived from Arabica and Robusta coffees in experimental coffee set. By assigning cutoff values in the S-plots, significant compounds were selected for p and p[corr] > |0.2|. Triangles indicate peaks detected by GC/MS. (From Jumhawan et al., *Journal of Agricultural and Food Chemistry,* 2013, 61: 7994–8001. With permission from ACS.)

The OPLS-DA score plot of Arabica coffee data sets is shown in Figure 7.7A. Discrimination between Kopi Luwak and regular coffee was obtained. The model was evaluated with R2 and Q2 values of 0.965 and 0.892, respectively. Interestingly, compounds that were uncorrelated with Kopi Luwak were quinic acid, caffeine, and caffeic acid. These compounds have been reported as contributors of bitterness as well as acidity in coffee.[36–39] In contrast, compounds that were predictive to

Kopi Luwak (i.e., over the cutoff value) included citric, malic, and glycolic acids. The OPLS-DA score plot of the Robusta coffee data sets (Figure 7.7C) was explained by R2 and Q2 values of 0.957 and 0.818, respectively. Caffeine, one of the bitter principles in coffee, was found to be significantly correlated with Robusta Kopi Luwak data sets. Robusta coffee has been reported to contain higher amounts of caffeine than Arabica. Thus, it tends to be bitter and flavorless, whereas Arabica coffee is considered to be milder, contain more aromatic compounds, and is more appreciated by the consumer.[45]

Candidates for discriminant markers for the authentication assessment of Arabica and Robusta coffees are listed in Table 7.4. The selected marker candidates met significant criteria in both OPLS-DA and SAM. Discriminant markers were chosen independently for the Arabica and Robusta coffees. To confirm whether these selected markers were generated as a result of animal digestion, we investigated cause–effect relationships by quantitating the discriminant marker candidates in green and roasted coffee beans from controlled processing, pre- and post-animal digestion. The results are displayed in Figure 7.8. In both the raw and roasted beans, citric acid was present in higher concentration after animal digestion, exhibiting a significant value difference ($p < 0.05$) between Kopi Luwak and regular coffee. The concentration of caffeine was also increased after digestion, but the difference was insignificant ($p > 0.05$). As a result of roasting, the glycolic acid concentration increased dramatically ($p < 0.001$) from 0.8 to 25–28 mg/kg. The production of aliphatic acids, including formic, acetic, glycolic, and lactic acids, has been reported during coffee roasting.[46] Therefore, among the selected marker candidates, we confirmed citric acid as a potential marker generated by animal digestion. Passage through the civet's gut, enriched with gastric juices and microbial activity, may have contributed to the increased levels of particular organic acids. Citric acid, malic acid, quinic acid, and chlorogenic acid are the main acids in coffee, and acidity is generated by complex reactions involving these organic acids during roasting.[37] Kopi Luwak has been reported to exhibit slightly higher acidity than regular coffee.[23] However, the correlation between the increased levels of particular acids as a result of animal digestion and

TABLE 7.4
Candidates for Discriminant Markers from OPLS-DA and SAM and Analytical Parameters for Quantitation

Discriminant Marker	RT (min)	VIP	RSD [%] (n = 3)		Linearity		LOD (mg/kg)	LOQ (mg/kg)
			RT	Area[a]	R^2	Range (µM)		
Glycolic acid	4.96	3.93	0.12	1.87	0.9999	1–1,000	0.021	0.066
Malic acid	9.05	5.53	0.05	2.29	0.9996	1–1,000	0.043	0.132
Pyroglutamic acid	9.43	1.7	0.05	3.36	0.9992	1–750	0.054	0.164
Citric acid	11.61	5.6	0.04	3.29	0.9997	1–1,000	0.504	1.526
Caffeine	12.18	2.28	0.04	3.51	0.9961	100–2,000	1.531	4.638
Inositol	13.45	4.47	0.03	5.09	0.9974	1–1,000	0.082	0.247

Source: Jumhawan et al., *Journal of Agricultural and Food Chemistry*, 2013, 61: 7994–8001. With permission from ACS.
[a] At 100 µM.

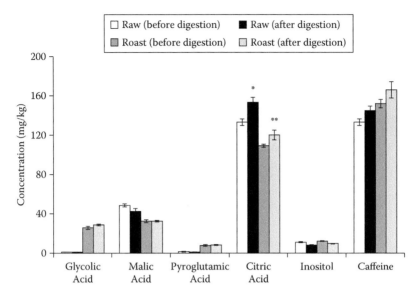

Figure 7.8 Concentration levels of six marker candidates from controlled processing. *, $p < 0.001$; **, $p < 0.05$. (From Jumhawan et al., *Journal of Agricultural and Food Chemistry,* 2013, 61: 7994–8001. With permission from ACS.)

the total acidity in coffee after roasting remains obscure and requires further investigation.

7.2.2.3 Validation of Applicability of Discriminant Markers for Authenticity Assessment

To verify the applicability of the selected marker candidates, we analyzed a validation coffee set that included authentic Kopi Luwak, commercial Kopi Luwak, commercial regular coffee, fake coffee, and coffee blend. With the exception of the authentic coffee, the remaining samples were purchased commercially. Generally, from harvest to preroasting, samples labeled "commercial Kopi Luwak" and "commercial regular coffee" were processed similarly to the corresponding coffees in the experimental set. However, in some cases, different roasting parameters were applied. Fake coffee was processed to approximate the sensory profile of Kopi Luwak.[24] Commercial regular coffees were selected from different production areas.

To examine the effectiveness of the selected markers in differentiating pure and coffee blends, we mixed two commercial

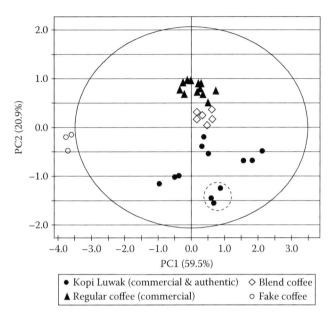

Figure 7.9 PCA score plot of validation coffee set. (From Jumhawan et al., *Journal of Agricultural and Food Chemistry,* 2013, 61: 7994–8001. With permission from ACS.)

digested coffees, Kopi Luwak Golden and Kopi Luwak Wahana, with a commercial regular coffee (Wahana regular) in a 50:50 (wt%) ratio. This would also compare the applicability of the discriminant markers when coffee beans from the same and different production areas were blended. Despite being selected independently, the six marker candidates were used together for method validation.

By subjecting all detected peaks to PCA, samples were populated into four clusters. The largest variance corresponded to fake coffee, as its results were clearly separated from others (data not shown). Next, we projected the six marker candidates as an inclusion list into the PCA to obtain an overview of their applicability toward sample differentiation. Similarly to the previous results, separation of the four coffee groups was observed. The PCA was explained by 59.5% and 20.9% variances in PC1 and PC2, respectively (Figure 7.9). Fake coffee was clustered away by PC1. Separation was likely because of attempts by the producer to obtain a profile similar to Kopi Luwak. In PC2, commercial Kopi Luwak, coffee blend, and commercial regular coffee could be differentiated. Both authentic and commercial Kopi Luwak were clustered within a close distribution area. Regardless of their origins and processing (roasting) parameters, commercial regular coffee data were populated in a close area, suggesting that these factors had the least significance for data separation. From the loading plot information, citric acid, malic acid, and inositol exhibited high contribution values for the Kopi Luwak data sets. Interestingly, these three marker candidates also showed the highest VIP values for constructing the discriminant model (Table 7.4).

To display the applicability of the selected discriminant markers in the differentiation of samples in the validation set, box plots were constructed using the relative peak intensities of citric acid, malic acid, and inositol. The box plots of malic acid and citric acid were able to differentiate commercial Kopi Luwak (Kopi Luwak Wahana), coffee blend, commercial regular coffee (Wahana regular), and fake coffee. However, the inositol box plot failed to differentiate these samples. Hence, we selected a double marker that employed an inositol–pyroglutamic acid ratio (Figure 7.10). Pyroglutamic acid was selected because it had the lowest contribution toward the separation of Kopi Luwak and regular coffee (data not shown).

We confirmed the ratio of the coffee blend by quantifying the discriminant marker constituents. The analytical parameters for quantitation are shown in Table 7.4. All authentic standards exhibited good linearity (0.99 or higher) and good repeatability for at least seven points in the applied concentration range in which analysis could be performed. To examine the quantitation validity, the LOD and LOQ for each discriminant marker were determined. The concentrations of the discriminant marker candidates in the coffee samples were determined to be higher than the LOD and LOQ of authentic standards. The concentration ratios of the selected markers, malic acid, citric acid, and the inositol–pyroglutamic acid ratio in all the sample blends ranged from 47.76% to 53.73%. This result showed a relatively low error in terms of the ratios of the discriminant markers in sample blends compared with their actual values. Moreover, the concentration of each discriminant

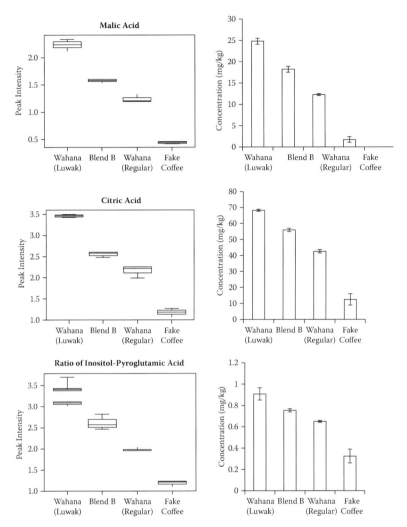

Figure 7.10 Box plots of peak intensity and concentrations of selected discriminant markers for validation. (From Jumhawan et al., *Journal of Agricultural and Food Chemistry,* 2013, 61: 7994–8001. With permission from ACS.)

marker corresponded well with the box plot constructed from its respective peak intensity (Figure 7.10). Hence, we confirmed the feasibility of using the proposed strategy for the robust authentication of coffee blend in a 50:50 (wt%) ratio.

In summary, to the best of our knowledge, this investigation represents the first attempt to address discriminant markers for the authentication of Kopi Luwak. Sample differentiation was greatly influenced by genetic diversity (coffee species), followed by decreasing contributions from animal perturbation and cultivation area. Because of the great variation among coffee species, candidates for the discriminant markers were selected independently for each species. The selected discriminant marker candidates were verified for the authentication of commercial coffee products. The proposed markers were able to differentiate commercial Kopi Luwak, commercial regular coffee, and fake coffee. In addition, at a certain ratio (50 wt% Kopi Luwak content), the feasibility of employing these discriminant markers to differentiate pure and mixed coffee was acceptable. Our findings highlighted the utility of metabolic profiling using GC/MS combined with multivariate analysis for the selection of discriminant markers for the authenticity assessments of valuable agricultural products. Discriminant markers are expected to perform as sole markers or in combination with sensory analysis by trained experts for the authentication of Kopi Luwak.

7.3 Metabolomics as a Tool for Informative Analysis and Relevance to Biological Science and Engineering

Case study: Metabolite profiles correlate closely with neurobehavioral function in experimental spinal cord injury in rats.[47]

7.3.1 Introduction

Spinal cord injury (SCI) often leads to severe disability due to permanent neurological impairment. The annual incidence of SCI lies between 10 to 80 persons per million and well over 30% of cases have some degree of tetraplegia.[48,49] The life expectancy of patients with acute SCI is dramatically and progressively shortened in relation to the degree of spinal cord injury and neurological deficit.[50] There are few therapeutic interventions that limit the extent of tissue damage following

SCI.[48] Consequently SCI is a catastrophic medical condition that dramatically reduces the patient's quality of life and imposes disproportionately large economic and social costs on affected individuals and society in general.[49]

The mechanistic insult to the spinal cord induces a variety of parallel pathophysiological processes.[51–53] In the early phase axons are disrupted, neural cell death occurs, and blood vessels are damaged, resulting in hemorrhage that exacerbates the ischemic neural injury. In the secondary phase necrotic death, electrolytic shifts, and edema continue. Cytotoxic levels of tissue debris, neurotransmitters, and reactive oxygen species are formed that promote an inflammatory response. Local tissue remodeling and neuroprotective and regenerative processes are also initiated that contribute to limited spontaneous improvement.

Accurate behavioral evaluation of SCI animals is an important tool for evaluating the therapeutic efficacy of pharmaceutical drugs. The Basso-Beattie-Bresnahan (BBB) locomotor rating scale is widely used to test behavioral consequences of spinal cord injury in rodents.[54] In this test joint movements, paw placement, weight support, and forelimb and hindlimb coordination are judged by experienced examiners according to the 21-point BBB locomotion scale. However, a simple and reliable quantitative chemical assay that readily could be adopted in a laboratory to monitor the outcome of SCI has yet to be developed. Furthermore, finding the biomarkers that could help to determine the degree of injury severity and to prognosticate neurologic recovery is one of the major challenges in management of SCI.[55,56] Identification of reliable, low-invasive biomarkers would be helpful for the clinician and patients in the choice of potential treatments.

Metabolomics is a new approach that involves the determination of changes in the levels of endogenous or exogenous metabolites in biological samples, owing to physiological stimuli or genetic modification.[57,58] The power of metabolomics lies in the global determination of metabolites, or patterns of biomarkers that increase or decrease as the result of a particular disease or injury. Here we have applied mass spectrometry (MS) based metabolomic technology in order to investigate the global metabolomic impact of SCI and to identify metabolites that could potentially be used to assess behavioral consequences of spinal cord injury.

7.3.2 Results

7.3.2.1 Metabolomic Analysis in Rat Spinal Cord Tissues After SCI or Sham Operation

To understand the metabolic impact of severe SCI, an untargeted metabolomic profiling approach was taken to assess the chemical milieu of the spinal cord. Female Sprague-Dawley rats were used in this experiment. They were divided into six groups: Two-, eleven- and thirty-days after injury and day-matched sham controls (without injury to the cord). Under the anesthesia by pentobarbital sodium, a laminectomy was performed at the ninth thoracic vertebra (T9), and the exposed spinal site. Then a 200-kdyn force was delivered to the exposed cord to produce a severe level of injury. Only rats that showed complete paralysis the day after the operation were used.

The motor function of the hindlimbs was evaluated according to the Basso-Beattie-Bresnahan locomotor scale[54] just before sampling. The BBB score ranged from 0 (complete paralysis) to 21 (normal gait). Two observers measured motor dysfunction of the right and left hindlimbs, respectively, and their mean value was employed as individual animal data. After the behavioral assessment, the rats were deeply anesthetized by pentobarbital sodium and 2.5 cm in length of spinal cord, centered at the lesion epicenter, were dissected to obtain enough amounts of tissue samples. Spinal cord tissues were immediately frozen by liquid nitrogen and stored under −70°C until measurement.

Mass spectrometry-based metabolomic profiling was performed as described in the literature.[59,60] Samples were homogenized and metabolites were extracted by the addition of cold methanol. The precipitated extract was split into four aliquots and dried under nitrogen and then in vacuo. The samples were resuspended in platform-specific solutions before they were applied to the instruments. The untargeted metabolomic profiling platform employed for this analysis was based on a combination of three independent platforms: ultrahigh performance liquid chromatography/tandem mass spectrometry (UHPLC/

MS/MS)[59] optimized for basic species, UHPLC/MS/MS optimized for acidic species, and gas chromatography/mass spectrometry (GC/MS).[60] Metabolites were identified by matching the
ions' chromatographic retention index and mass spectral fragmentation signatures with reference library entries created from
authentic standard metabolites. For ions that were not covered
by the standards, additional library entries were added based on
their unique retention time and ion signatures. Peak ion counts
for each compound in each sample were used for statistical analysis, resulting in the comparisons of relative concentrations.

There were 405 metabolites, detected of which 283 matched
known structures in the chemical reference library. SCI
induced dramatic metabolic changes with more than 25% of
the profiled metabolites significantly altered. Clear metabolic
separation between the sham and SCI animals was observed
by principal component analysis (Figure 7.11). Moreover, time-
dependent changes in metabolite profiles after SCI operation
were also observed reflecting the dramatic longitudinal effect
of the damage. Typical metabolites contributing to the separation in PCA score plot are listed in Table 7.5.

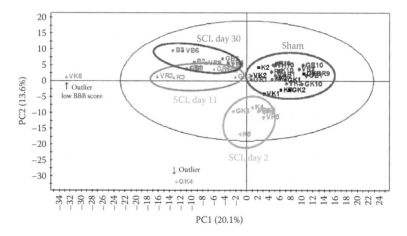

Figure 7.11 PCA Score plot of rat spinal cord samples in SCI or Sham-
operated animals with identification number. Black ellipse in the score plot
illustrates the 95% confidence regions. PCA score plot displays the distinct dis-
similarities between Sham animals and SCI animals, and also suggests the
time-dependent changes in metabolite profiles after SCI operation. There are
two outliers in the PCA score plot; one of them (animal No. VK6) in SCI day11
group showed low BBB score. (From Fujieda et al., *PloS ONE*, 2012, 7, e43152.)

TABLE 7.5
Altered Spinal Cord Metabolites in the Primary and Secondary Phase When Contrasting SCI and Sham-Treated Rats[a,b]

Early Metabolic Phase	SCI Versus Sham					
Metabolite	Fold Change			p-Value		
Lipid Mediator	Day 2	Day 11	Day 30	Day 2	Day 11	Day 30
oleic ethanolamide	**7.86**	1.15	1.06	0.00	0.32	0.51
palmitoyl ethanolamide	**7.09**	1.06	1.04	<0.01	0.69	0.70
prostaglandin E2	**1.32**	1.08	NA	0.04	0.18	—
Polyamine						
ornithine	**1.77**	0.96	0.79	0.02	0.60	0.30
putrescine	**2.39**	**2.21**	1.31	<0.01	<0.01	0.14
spermidine	0.94	1.04	0.96	0.44	0.78	0.63
5-methylthioadenosine	0.92	0.95	0.87	0.37	0.53	0.10
Neurotransmitters						
glutamate	**0.78**	**0.90**	**0.87**	<0.01	0.01	<0.01
glutamine	**0.94**	1.04	**1.10**	0.05	0.32	0.02
N-acetyl-aspartyl-glutamate	0.86	**0.80**	**0.74**	0.06	<0.01	<0.01
gamma-aminobutyrate	0.86	0.90	**0.78**	0.06	0.44	0.01
aspartate	**0.65**	**0.85**	**0.91**	0.00	0.00	0.04
N-acetylaspartate	0.83	**0.63**	**0.54**	0.31	0.00	<0.01
Secondary Metabolic Phase	SCI Versus Sham					
Metabolite	Fold Change			p-Value		
Membrane remodeling	Day 2	Day 11	Day 30	Day 2	Day 11	Day 30
linolenate (18:3n3 or 6)	**1.45**	1.87	1.21	0.01	0.24	0.12
dihomo-linolenate (20:3n3 or n6)	0.91	**1.60**	**1.74**	0.37	0.00	<0.01
eicosapentaenoate (20:5n3)	**1.41**	**2.25**	**2.47**	0.01	<0.01	<0.01
docosapentaenoate (22:5n3)	**1.94**	**3.02**	**2.87**	<0.01	<0.01	<0.01

(Continued)

TABLE 7.5 (CONTINUED)
Altered Spinal Cord Metabolites in the Primary and Secondary Phase When Contrasting SCI and Sham-Treated Rats[a,b]

Secondary Metabolic Phase	SCI Versus Sham					
Metabolite	Fold Change			p-Value		
Membrane remodeling	Day 2	Day 11	Day 30	Day 2	Day 11	Day 30
docosapentaenoate (22:5n6)	**1.61**	**2.84**	**4.95**	0.00	0.00	<0.01
docosahexaenoate (22:6n3)	**1.89**	**2.00**	**2.19**	0.01	<0.01	<0.01
glycerophosphorylcholine	0.92	**1.71**	**1.79**	0.39	<0.01	<0.01
docosadienoate (22:2n6)	**1.28**	**2.31**	**2.14**	0.04	<0.01	<0.01
docosatrienoate (22:3n3)	**1.79**	**3.34**	**2.79**	0.00	<0.01	<0.01
adrenate (22:4n6)	**1.25**	**2.01**	**1.94**	0.03	<0.01	<0.01
1-palmitoyl-GPI	1.00	**1.37**	**1.57**	0.95	0.00	<0.01
1-stearoyl-GPI	1.12	**1.19**	**1.28**	0.39	0.05	<0.01
1-arachidonoyl-GPI	1.04	**1.84**	**1.66**	0.68	0.01	0.02
1-oleoyl-GPS	**2.24**	**2.24**	**2.76**	<0.01	<0.01	<0.01
2-oleoyl-GPS	1.20	**1.72**	**2.11**	0.25	0.02	<0.01
ethanolamine	0.97	**1.71**	1.17	0.60	0.02	0.23
phosphoethanolamine	0.79	**1.79**	**1.38**	0.05	<0.01	<0.01
Antioxidant Defense						
ascorbate	**0.61**	**1.42**	1.11	<0.01	0.01	0.19
glutathione, (GSSG)	0.85	**3.79**	1.68	0.77	0.00	0.20
alpha-tocopherol	1.02	**1.30**	**1.29**	0.72	0.02	0.03
ergothioneine	1.07	**1.42**	**1.25**	0.34	<0.01	0.01

Source: Fujieda et al., *PloS ONE*, 2012, 7, e43152.

[a] GPI refers to glycerophosphoinositol and GPS refers to glycerophosphoserine.

[b] Statistically significant changes are in bold ($p < 0.05$, Welch's two sample t test).

Close examination of the data revealed two metabolic phases. In the early metabolic phase (day 2) lipid mediators, polyamines, and neurotransmitter-related compounds were strongly altered (Table 7.5). An increase in lipid mediators, oleic ethanolamide (OEA), palmitoyl ethanolamide (PEA), and prostaglandin E2 (PGE_2) were observed at day 2. The subsequent return to basal levels indicated that their elevation was a direct response to the injury. Polyamines were also strongly affected. In particular, putrescine was greatly elevated by day two and returned to basal level after day 11 consistent with reports that ornithine decarboxylase, the rate-limiting enzyme in polyamine synthesis, is strongly activated in response to stress.[61,62] This activation may indicate both secondary pathogenesis and induction of the neuroprotective response. Dramatic changes in neurotransmitters and several of their precursors were observed at day 2 that proceeded until day 30 following SCI. Many of these neurotransmitters were significantly decreased, possibly reflecting neuronal cell death and decreased synthesis capacity after the injury.

In a secondary metabolic phase (day 11 and day 30), metabolites related to membrane remodeling or antioxidant defense were significantly altered (Table 7.5). Increased levels of the phospholipid degradation products glycerophosphocholine, lysolipids, and fatty acids were observed. Several of the elevated fatty acids, including dihomo-linolenate, docosapentaenoate (DPA), and docosahexaenoate (DHA), are precursors to pro- and anti-inflammatory inflammation molecules. This biochemical signature is indicative of a dynamic inflammatory environment following SCI. Increased levels of ethanolamine and phosphoethanolamine, intermediates in phospholipid synthesis, implied that this pathway was also elevated from day 11. The activation of both pathways suggested that the inflammatory response following SCI removed debris from the injury and also initiated the repair process.

Oxidative stress is thought to be a major contributor to the damage that occurs in the spinal cord following injury due to secondary effects. Increased levels of antioxidants such as ergothioneine, α-tocopherol, glutathione, and ascorbate were observed from day 11 following SCI suggesting that a broad spectrum of the defense system was mobilized.

7.3.2.2 Metabolite Profiles Correlate Closely with BBB Scores

The 21-point open field locomotion scale developed by Basso, Beattie, and Bresnahan is widely used to assess locomotive recovery of SCI in rodents. BBB scores were measured after severe SCI (200-kdyn impact force) at days 11 and 30. Severe loss of locomotive activity was followed by limited spontaneous improvement in the BBB score over the 30 days, consistent with previous reports.[54]

The distinct metabolic changes observed as a result of SCI raised the possibility that they may reflect the locomotive recovery process. The metabolite levels to the BBB locomotive scores at both time points were compared. Many metabolites were observed to correlate greatly with the BBB score. Among the top metabolites were compounds associated with the biochemical events following SCl, including neurotransmitter-related compounds and lipids (Figure 7.12). Decreased levels of *N*-acetyl-aspartyl-glutamate (NAAG) and *N*-acetyl-aspartate (NAA) resulted in positive correlations with the BBB score. In contrast, increased levels of several lipids including DHA and ω3-DPA (n3 DPA) showed highly negative correlations. These observations indicated that there was a biochemical basis for the BBB score suggesting that these metabolites may be potential chemical markers of locomotive recovery.

7.3.2.3 Correlation of NAA with BBB Score

In order to illustrate that single metabolites can be used to predict the locomotive score, we developed a GC/MS-based assay to measure the absolute spinal cord concentrations of NAA in independent cohorts of mild to severe spinal cord injured rats. We focused on NAA because it is an abundant neurotransmitter-related compound that has been studied in SCI models and can be measured noninvasively by proton magnetic resonance spectroscopy (^{1}H-MRS).[63,64]

Eight-week-old female Sprague-Dawley rats were purchased and used within a week. We chose the three injury forces to get animals with a variety of BBB scores. They were divided into four groups of: Sham controls ($n = 2$) and 100-kdyn (mild: $n = 6$), 150-kdyn (moderate: $n = 6$), and 200-kdyn (severe: $n = 6$) injured group. Animals were prepared as described earlier and 100-, 150-, or 200-kdyn injury-inducing forces were delivered to the exposed cord to produce a wide variety of symptoms. Only animals that showed complete paralysis (BBB score = 0) the day after the operation were used. The motor function of the hindlimbs was evaluated according to the BBB scale just before the sampling (at 32–34 days postinjury). After the behavioral assessment, the rats were deeply anesthetized and spinal cord tissues around the epicenter were collected as described in the previous section. Spinal cord tissues were homogenized and the aqueous metabolites were extracted by a modified Bligh-Dyer method.[65] For the stable isotope dilution, NAA-^{13}C was added to the specimens as an internal standard. Samples were converted to their trimethylsilyl derivatives (TMS) prior to GC/MS analysis and NAA concentration in spinal cord tissue was quantitated by GC/MS/MS.

As illustrated in Figure 7.13, the NAA concentrations in the spinal cord of injured animals were lower than those of Sham animals and correlated greatly with the BBB score ($R > 0.9$). This result validated the findings from the metabolomics study and suggested that NAA could be used to predict the locomotive function.

7.3.3 Discussion

In this study we successfully applied biochemical profiling to assess the global metabolic changes associated with SCI. Dramatic metabolic changes were observed as a result of SCI. These changes were predominantly reflective of pathophysiological processes such as inflammation, tissue damage, clearance, and remodeling after SCI, and were quite consistent with previous findings.[51–53] In addition, we found several metabolites, including NAA, NAAG, and the ω-3 fatty acids

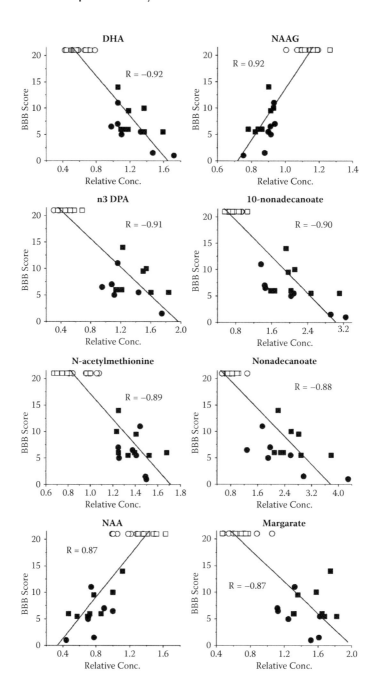

highly correlated with the established BBB score. Although Kwo et al.[66] have reported potassium and calcium in the spinal cord were correlated to injury severity, this is the first report that shows the correlation among the BBB scoring system, the index of neurobehavioral function, and metabolic changes in a rodent model. Moreover, we could find a biomarker candidate, NAA, which could be measured noninvasively in humans and help to determine the degree of injury severity and to prognosticate neurologic recovery after SCI from this study.

Several microarray and proteomics studies were conducted to characterize the gene or protein expressions in spinal cord after experimental SCI in rats. Carmel et al.[67] showed the neuronal loss and inflammatory response in gene expressions in the early phase. Resnick et al.[68] analyzed the delayed response in gene expressions and reported that the repair process was observed in spinal cord tissues. Yan et al.[69] investigated the time courses of protein expressions after the contusive SCI and reported the alterations of proteins associated with apoptosis, metabolism, and cytoskeleton organization. The changes in metabolites in spinal cord we demonstrated were also consistent with other omics profiles after SCI.

In the early metabolic phase, a complex picture of the inflammatory response emerged. The lipid mediators, PGE_2 and PEA, were induced shortly after the injury. PGE_2 has proinflammatory effects that may play a role in maintaining neuropathic pain.[70,71] In contrast, PEA has been shown to have anti-inflammatory properties indicating that a complex signaling network may have been activated to prevent an overactive and damaging inflammatory response.[72] Consistent with this notion, ω-6 and ω-3 polyunsaturated fatty acids, precursors of pro- and anti-inflammatory lipid mediators, were increasingly

Figure 7.12 Correlation plots comparing BBB scores with metabolite levels at days 11 (circle) and 30 (square). The eight named metabolites with the highest $|R|$ score are displayed (see Table 7.5 for complete list). Sham animals (day 11 and 30) received full BBB score (=21), consistent with normal movement. SCI at day 11 and 30 are depicted as ●($n = 7$) and ■($n = 8$). Shams at day 11 and 30 are displayed as ○($n = 8$) and □($n = 8$). All data points of Sham animals and SCI animals were used to calculate the R values. (From Fujieda et al., *PloS ONE*, 2012, 7, e43152.)

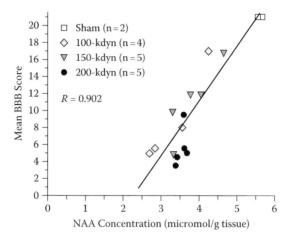

Figure 7.13 Correlation between absolute NAA concentration in the spinal cord and the corresponding BBB scores 32–34 days after SCI. SCI was induced by 100-, 150-, or 200-kdyn impact force. Sham animals received full BBB score (=21), consistent with normal movement. (From Fujieda et al., *PloS ONE*, 2012, 7, e43152.)

elevated throughout the experiment suggesting that the inflammatory response was continuously being modulated. Activation of the inflammatory response is strongly associated with ROS formation. A broad spectrum of antioxidants was elevated consistent with mobilization of the defense system in response to the damage and secondary injury. These findings indicate that inflammation and oxidative stress pathways are active for a prolonged time period. It would thus be expected that they strongly contribute to the pathological outcome.

Neurotransmitters and related precursors (NAAG, NAA, aspartate, and glutamate) remained suppressed following SCI. NAAG, NAA, and glutamate are abundant within neuronal cells, suggesting that the observed changes may reflect neuronal cell death.[73] Rapid release of excessive glutamate and other neurotransmitter-related compounds that may contribute directly to cellular damage has been observed following SCI.[74,75] The inability to restore these levels suggested that neuronal regeneration following SCI was limited.

We showed that many of the dramatic metabolic changes associated with SCI closely correlated with the established BBB locomotive score. The 21-point BBB scale represents a detailed and ordinal categorization of hindlimb locomotor recovery after spinal cord injury. In the BBB scoring system, joint movements are mainly assessed in the low score range (score 0–7), weight-supported plantar placements are evaluated in the middle score range (score 7–12), and coordination of movement is scored in the high score range (score 13–21).[54] Metabolites that correlated best with the BBB scores were strongly associated with the biochemical events following SCI, including neurotransmitter-related compounds and lipids. The most abundant peptide neurotransmitter in the mammalian nervous system, NAAG, correlated highly with the BBB score. NAAG has been shown to be an important marker of neuronal viability. NAA levels are closely associated with neuronal function,[76] and NAA concentration in the spinal cord was decreased rapidly after the spinal cord injury in rats.[77] Decreased NAA levels have also been reported in various neurodegenerative diseases such as Alzheimer's disease, amyotrophic lateral sclerosis, and Parkinson's disease.[78–80] Blamire et al.[81] showed that decreases in NAA closely correlated with neuronal or axonal dysfunction loss in multiple sclerosis. The NAAG and NAA levels were indicative of initial neuronal damage followed by limited regeneration, possibly reflecting impaired synthesis capacity.

Several fatty acids correlated greatly with the BBB score. Many of these are important membrane constituents when incorporated into phospholipids. Their changes are reflective of damage and regeneration of the membrane. Increased levels of two abundant ω-3 fatty polyunsaturated acids (ω-3 PUFA), DHA and 5n3-DPA, showed strong negative correlations with the neurobehavioral function. These ω-3 PUFAs are specifically enriched in the brain and mainly anchored in the neuronal membrane as phospholipids, where they are involved in the maintenance of normal neurological function.[82] These increases in ω-3 PUFAs after SCI might reflect the cell damage in the spinal cord. On the other hand, recent evidence shows that ω-3 fatty acids or DHA can modulate several of the processes that contribute to secondary degeneration in the CNS and improve recovery from SCI.[83,84] It is presumable that a DHA increase in advance of SCI might promote neuroprotection by triggering the release of DHA-derived mediators, such as resolvins or neuroprotectin D1.[85] In addition, multiple double bonds of ω-3 PUFAs are excellent targets for lipid peroxidation that could potentiate neurotoxicity; the increases of ω-3 PUFAs may also function as a free-radical scavenger to reduce the neuronal oxidative stress after SCI. The strong correlation with these metabolites suggests that the BBB score is closely related to the metabolic basis for cell damage, regeneration, and inflammatory response.

We developed a targeted assay to measure NAA quantitatively and conducted an additional experiment to demonstrate the utility of NAA measurement for the prediction of neurobehavioral function after SCI. High correlation with the BBB score was observed (Figure 7.13) validating the findings from the metabolomics analysis. In this experiment, because many animals were severely injured and 6 of 14 showed almost the same BBB scores (4.5–5.5), it was not clear whether the NAA concentrations in severely injured animals highly correlated with BBB scores. However, the overall correlation was very high ($R > 0.9$), and a single metabolite, NAA, appears to correlate with and reflect what the locomotive score is likely to be. *N*-acetyl-aspartate is a free amino acid, almost exclusively located in neurons and axons that can be measured noninvasively by [1]H-magnetic resonance spectroscopy.[63,64] Using ^{1}H-MRS, Qian and his colleagues[64] showed that NAA decreased rapidly after SCI and stayed depressed for 56 days in the epicenter as well as the rostral and caudal segments in rats, consistent with the findings in this study. Our results suggest that the neurobehavioral function after SCI can be estimated by measuring the NAA concentration around the epicenter in the spinal cord. Because NAA concentration in human spinal cord could be measured noninvasively by ^{1}H-MRS,[86,87] NAA levels in spinal cord might provide a meaningful biomarker that could help to determine the degree of injury severity and to prognosticate neurologic recovery.

References

1. Yoshida, R. et al., Metabolomics-based systematic prediction of yeast lifespan and its application for semi-rational screening of ageing-related mutants, *Aging Cell*, 2010, 9: 616–625.
2. Bader, G. D. et al., Functional genomics and proteomics: Charting a multidimensional map of the yeast cell, *Trends in Cell Biology*, 2003, 13: 344–356.
3. Shilling, P. D. and Kelsoe, J. R., Functional genomics approaches to understanding brain disorders, *Pharmacogenomics*, 2002, 3: 31–45.
4. The European Bioinformatics Institute, What is functional genomics? 2014, http://www.ebi.ac.uk/training/online/course/functional-genomics-introduction-ebi-resources/what-functional-genomics.
5. Lafaye, A. et al., Combined proteome and metabolite-profiling analyses reveal surprising insights into yeast sulfur metabolism, *Journal of Biological Chemistry*, 2005, 280: 24723–24730.
6. Pir, P. et al., Integrative investigation of metabolic and transcriptomic data, *BMC Bioinformatics*, 2006, 7: 203.
7. Raamsdonk, L. M., et al., A functional genomics strategy that uses metabolome data to reveal the phenotype of silent mutations, *Nature Biotechnology*, 2001, 19: 45–50.
8. Allen, J., et al. High-throughput classification of yeast mutants for functional genomics using metabolic footprinting, *Nature Biotechnology*, 2003, 21: 692–696.
9. Soga, T. et al., Differential metabolomics reveals ophthalmic acid as an oxidative stress biomarker indicating hepatic glutathione consumption, *Journal of Biological Chemistry*, 2006, 281: 16768–16776.
10. Hirayama, A., et al., Quantitative metabolome profiling of colon and stomach cancer microenvironment by capillary electrophoresis time-of-flight mass spectrometry, *Cancer Research*, 2009, 69: 4918–4925.
11. Serkova, N. J. and Glunde, K., Metabolomics of cancer, *Methods in Molecular Biology*, 2009, 520: 273–295.
12. Mortimer, R. K. and Johnston, J. R., Life span of individual yeast cells, *Nature*, 1959, 183: 1751–1752.
13. Kaeberlein, M. et al., Genes determining yeast replicative life span in a long-lived genetic background, *Mechanisms of Ageing Development*, 2005, 126: 491–504.

14. Kaeberlein, M., et al. Regulation of yeast replicative life span by TOR and Sch9 in response to nutrients, *Science,* 2005, 310: 1193–1196.

15. Piper, P. W. Long-lived yeast as a model for ageing research, *Yeast,* 2006, 23: 215–226.

16. Jiang, J. C., et al. An intervention resembling caloric restriction prolongs life span and retards aging in yeast, *FASEB Journal,* 2000, 14: 2135–2137.

17. Lin, S. J., et al. Calorie restriction extends *Saccharomyces cerevisiae* lifespan by increasing respiration, *Nature,* 2002, 418: 344–348.

18. Imai, S. et al., Transcriptional silencing and longevity protein Sir2 is an NAD-dependent histone deacetylase, *Nature,* 2000, 403: 795–800.

19. Lindstrom, D. L. and Gottschling, D. E., The mother enrichment program: A genetic system for facile replicative lifespan analysis in *Saccharomyces cerevisiae, Genetics,* 2009, 183: 413–422.

20. Jumhawan, U. et al., Selection of discriminant marker candidates for authentication of Asian palm civet coffee (Kopi Luwak): A metabolomics approach, *Journal of Agricultural and Food Chemistry,* 2013, 61: 7994–8001.

21. Onishi, N., From dung to coffee brew with no aftertaste, 2013, http://www.nytimes.com/2010/04/18/world/asia/18civetcoffee.html

22. Joshi, A. R. et al., Influence of food distribution and predation pressure on spacing behavior in palm civets, *Journal of Mammalogy,* 1995, 76: 1205–1212.

23. Marcone, M. F., Composition and properties of Indonesian palm civet coffee (Kopi Luwak) and Ethiopian civet coffee, *Food Research International*, 2004, 37: 901–912.

24. White Koffie first low acid coffee, 2013, http://blog.kopiluwak.org/white-koffie-first-low-acid-coffee

25. Blow, N., Metabolomics: Biochemistry's new look, *Nature,* 2008, 455: 697–700.

26. Choi, M. et al., Determination of coffee origins by integrated metabolomics approach of combining multiple analytical data, *Food Chemistry,* 2010, 121: 1260–1268.

27. Wei, F., et al., ^{13}C NMR-based metabolomics for the classification of green coffee beans according to variety and origin, *Journal of Agricultural and Food Chemistry,* 2012, 60: 10118–10125.

28. Fujimura, Y. et al., Metabolomics-driven nutraceutical evaluation of diverse green tea cultivars, *PloS ONE*, 2011, 6: e23426.

29. Pongsuwan, W. et al., Prediction of Japanese green tea ranking by gas chromatography/mass spectrometry-based hydrophilic metabolite profiling, *Journal of Agricultural and Food Chemistry*, 2007, 55: 231–236.
30. Jonsson, P. et al., Predictive metabolite profiling applying hierarchical multivariate curve resolution to GC-MS data—a potential tool for multi-parametric diagnosis, *Journal of Proteome Research,* 2006, 5: 1407–1414.
31. Wiklund, S. et al., Visualization of GC/TOF-MS-based metabolomics data for identification of biochemically interesting compounds using OPLS class models, *Analytical Chemistry,* 2008, 80: 115–122.
32. Tusher, V. G. et al., Significance analysis of microarrays applied to the ionizing radiation response, *Proceedings of the National Academy of Sciences USA,* 2001, 98: 5116–5121.
33. Dalluge, J. et al., Resistively heated gas chromatography coupled to quadrupole mass spectrometry, *Journal of Separation Science,* 2005, 25, 608–614.
34. Shimadzu, Inc., GCMS-QP2010 Ultra, 2013, http://www.shimadzu.com/an/gcms/qpultra.html
35. Wermelinger, T. et al., Quantification of the Robusta fraction in a coffee blend via Raman spectroscopy: Proof of principle, *Journal of Agricultural and Food Chemistry,* 2011, 59: 9074–9079.
36. Moon, J. K. et al., Role of roasting conditions in the level of chlorogenic acid content in coffee beans: Correlation with coffee acidity, *Journal of Agricultural and Food Chemistry,* 2009, 57: 5365–5369.
37. Rodrigues, C. et al., Application of solid-phase extraction to brewed coffee caffeine and organic acid determination by UV/HPLC, *Journal of Food Composition and Analysis,* 2007, 20: 440–448.
38. Blumberg, S. et al., Quantitative studies on the influence of the bean roasting parameters and hot water percolation on the concentrations of bitter compounds in coffee brew, *Journal of Agricultural and Food Chemistry,* 2010, 58: 3720–3728.
39. Perrone, D. et al., Fast simultaneous analysis of caffeine, trigonelline, nicotinic acid, and sucrose in coffee by liquid chromatography–mass spectrometry, *Food Chemistry,* 2008, 110: 1030–1035.
40. Wei, F. et al., Roasting process of coffee beans as studied by nuclear magnetic resonance: Time course of changes in composition, *Journal of Agricultural and Food Chemistry,* 2012, 60: 1005–1012.

41. Liberto, E. et al., Non-separative headspace solid phase micro-extraction-mass spectrometry profile as a marker to monitor coffee roasting degree, *Journal of Agricultural and Food Chemistry*, 2013, 61: 1652–1660.

42. Ruosi, M. R. et al., A further tool to monitor the coffee roasting process: Aroma composition and chemical indices, *Journal of Agricultural and Food Chemistry*, 2012, 60: 11283–11291.

43. Lundstedt, T. et al., Experimental design and optimization, *Chemometrics and Intelligent Laboratory Systems*, 1998, 42: 3–40.

44. Erikkson, L. et al., Methods for reliability and uncertainty assessment and for applicability evaluations of classification- and regression-based QSARs, *Environmental Health Perspectives*, 2003, 111: 1361–1375.

45. Wintgens, J. N., Botany, genetics and genomics of coffee. In *Coffee: Growing, Processing, Sustainable Production a Guidebook for Growers, Processors, Traders and Researchers*, second revised edition, Vol. 2, 2009, New York: Wiley VCH, pp. 25–60.

46. Ginz, M. et al., Formation of aliphatic acids by carbohydrate degradation during roasting of coffee, *European Food Research and Technology*, 2000, 211: 404–410.

47. Fujieda, Y. et al., Metabolite profiles correlate closely with neurobehavioral function in experimental spinal cord injury in rats, *PloS ONE*, 2012, 7, e43152.

48. McDonald, J. W. and Sadowsky, C., Spinal-cord injury, *Lancet*, 2002, 359: 417–425.

49. Wyndaele, M. and Wyndaele, J. J., Incidence, prevalence and epidemiology of spinal cord injury: What learns a worldwide literature survey? *Spinal Cord*, 2006, 44: 523–529.

50. Hulsebosch, C. E., Recent advances in pathophysiology and treatment of spinal cord injury, *Advances in Physiology Education*, 2002, 26: 238–255.

51. Fitch, M. T. and Silver, J., CNS injury, glial scars, and inflammation: Inhibitory extracellular matrices and regeneration failure, *Experimental Neurology*, 2008, 209: 294–301.

52. Tator, C. H., Experimental and clinical studies of the pathophysiology and management of acute spinal cord injury, *Journal of Spinal Cord Medicine*, 1996, 19: 206–214.

53. Tator, C. H., Biology of neurological recovery and functional restoration after spinal cord injury, *Neurosurgery*, 1998, 42: 696–707, discussion 707–698.

54. Basso, D.M. et al., A sensitive and reliable locomotor rating scale for open field testing in rats, *Journal of Neurotrauma*, 1995, 12: 1–21.

55. Pouw, M. H. et al., Biomarkers in spinal cord injury, *Spinal Cord,* 2009, 47: 519–525.
56. Lubieniecka, J. M., et al., Biomarkers for severity of spinal cord injury in the cerebrospinal fluid of rats, *PloS ONE,* 2011, 6: e19247.
57. Kaddurah-Daouk, R. et al., Metabolomics: A global biochemical approach to drug response and disease, *Annual Review of Pharmacology and Toxicology,* 2008, 48: 653–683.
58. Lindon, J. C. et al., Metabonomics in pharmaceutical R&D, *FEBS Journal,* 2007, 274: 1140–1151.
59. Evans, A. M. et al., Integrated, nontargeted ultrahigh performance liquid chromatography/electrospray ionization tandem mass spectrometry platform for the identification and relative quantification of the small-molecule complement of biological systems, *Analytical Chemistry,* 2009, 81: 6656–6667.
60. Sha, W. et al., Metabolomic profiling can predict which humans will develop liver dysfunction when deprived of dietary choline, *FASEB Journal,* 2010, 24: 2962–2975.
61. Mautes, A. E. et al., Changes in ornithine decarboxylase activity and putrescine concentrations after spinal cord compression injury in the rat, *Neuroscience Letters,* 1999, 264: 153–156.
62. Dienel, G. A. and Cruz, N. F., Induction of brain ornithine decarboxylase during recovery from metabolic, mechanical, thermal, or chemical injury, *Journal of Neurochemistry,* 1984, 42: 1053–1061.
63. Erschbamer, M. et al., 1H-MRS in spinal cord injury: Acute and chronic metabolite alterations in rat brain and lumbar spinal cord, *European Journal of Neuroscience,* 2011, 33: 678–688.
64. Qian, J. et al., Neuronal and axonal degeneration in experimental spinal cord injury: *In vivo* proton magnetic resonance spectroscopy and histology, *Journal of Neurotrauma,* 2010, 27: 599–610.
65. Hayashi, S. et al., A novel application of metabolomics in vertebrate development, *Biochemical and Biophysics Research Communications,* 2009, 386: 268–272.
66. Kwo, S. et al., Spinal cord sodium, potassium, calcium, and water concentration changes in rats after graded contusion injury, *Journal of Neurotrauma,* 1989, 6: 13–24.
67. Carmel, J. B. et al., Gene expression profiling of acute spinal cord injury reveals spreading inflammatory signals and neuron loss, *Physiological Genomics,* 2001,7: 201–213.
68. Resnick, D. K. et al., Molecular evidence of repair and plasticity following spinal cord injury, *Neuroreport,* 2004, 15: 837–839.

69. Yan, X. et al., Proteomic profiling of proteins in rat spinal cord induced by contusion injury, *Neurochemistry International,* 2010, 56: 971–983.

70. Abe, T. et al., Fyn kinase-mediated phosphorylation of NMDA receptor NR2B subunit at Tyr1472 is essential for maintenance of neuropathic pain, *European Journal of Neuroscience,* 2005, 22: 1445–1454.

71. Ma, W., Chabot, J. G., Vercauteren, F., and Quirion, R., Injured nerve-derived COX2/PGE2 contributes to the maintenance of neuropathic pain in aged rats, *Neurobiolic Aging,* 2010, 31: 1227–1237.

72. Genovese, T. et al., Effects of palmitoylethanolamide on signaling pathways implicated in the development of spinal cord injury, *Journal of Pharmacology and Experimental Therapeutics,* 2008, 326: 12–23.

73. Sager, T. N. et al., Changes in N-acetyl-aspartate content during focal and global brain ischemia of the rat, *Journal of Cerebral Blood Flow and Metabolism,* 1995, 15: 639–646.

74. Faden, A. I. et al., The role of excitatory amino acids and NMDA receptors in traumatic brain injury, *Science,* 1989, 244: 798–800.

75. Vera-Portocarrero, L. P. et al., Rapid changes in expression of glutamate transporters after spinal cord injury, *Brain Research,* 2002, 927: 104–110.

76. Demougeot, C. et al., N-acetylaspartate: A literature review of animal research on brain ischaemia, *Journal of Neurochemistry,* 2004, 90: 776–783.

77. Falconer, J. C. et al., Time dependence of N-acetyl-aspartate, lactate, and pyruvate concentrations following spinal cord injury, *Journal of Neurochemistry,* 1996, 66: 717–722.

78. Griffith, H. R. et al., Brain N-acetylaspartate is reduced in Parkinson disease with dementia, *Alzheimer Disease and Associated Disordors,* 2008, 22: 54–60.

79. Giroud, M. et al., Reduced brain N-acetyl-aspartate in frontal lobes suggests neuronal loss in patients with amyotrophic lateral sclerosis, *Neurologic Research,* 1996, 18: 241–243.

80. Schott, J. M. et al., Short echo time proton magnetic resonance spectroscopy in Alzheimer's disease: A longitudinal multiple time point study, *Brain,* 2010, 133: 3315–3322.

81. Blamire, A. M. et al., Axonal damage in the spinal cord of multiple sclerosis patients detected by magnetic resonance spectroscopy, *Magnetic Resonance in Medicine,* 2007, 58: 880–885.

82. Su, H. M., Mechanisms of n-3 fatty acid-mediated development and maintenance of learning memory performance, *Journal of Nutritional Biochemistry,* 2010, 21: 364–373.

83. King, V. R. et al., Omega-3 fatty acids improve recovery, whereas omega-6 fatty acids worsen outcome, after spinal cord injury in the adult rat, *Journal of Neuroscience,* 2006, 26: 4672–4680.

84. Figueroa, J. D. et al., Docosahexaenoic acid pretreatment confers protection and functional improvements after acute spinal cord injury in adult rats, *Journal of Neurotrauma,* 2012, 29: 551–566.

85. Bazan, N. G., The onset of brain injury and neurodegeneration triggers the synthesis of docosanoid neuroprotective signaling, *Cellular and Molecular Neurobiology,* 2006, 26: 901–913.

86. Kendi, A. T. et al., MR spectroscopy of cervical spinal cord in patients with multiple sclerosis, *Neuroradiology,* 2004, 46: 764–769.

87. Marliani, A. F. Clementi, V., Albini-Riccioli, L., Agati, R., and Leonardi, M., Quantitative proton magnetic resonance spectroscopy of the human cervical spinal cord at 3 Tesla, *Magnetic Resonance in Medicine,* 2007, 57: 160–163.

Index

D

E

F

G

H